Porous Metals: Preparation, Microstructure, Properties and Performance

Porous Metals: Preparation, Microstructure, Properties and Performance

Guest Editors

Yuyuan Zhao
Huiping Tang

MDPI

Basel • Beijing • Wuhan • Barcelona • Belgrade • Novi Sad • Cluj • Manchester

Guest Editors

Yuyuan Zhao
School of Mechanical and
Automotive Engineering
Ningbo University of
Technology
Ningbo
China

Huiping Tang
Advanced Materials Additive
Manufacturing Innovation
Research Center
Hangzhou City University
Hangzhou
China

Editorial Office
MDPI AG
Grosspeteranlage 5
4052 Basel, Switzerland

This is a reprint of the Special Issue, published open access by the journal *Materials* (ISSN 1996-1944), freely accessible at: www.mdpi.com/journal/materials/special_issues/1YQ475552J.

For citation purposes, cite each article independently as indicated on the article page online and using the guide below:

Lastname, A.A.; Lastname, B.B. Article Title. *Journal Name* **Year**, *Volume Number*, Page Range.

ISBN 978-3-7258-4070-0 (Hbk)
ISBN 978-3-7258-4069-4 (PDF)
https://doi.org/10.3390/books978-3-7258-4069-4

Contents

Preface

Porous metals (also known as metal foams and cellular metals) are a special class of composite materials composed of a metal phase and a gaseous phase. The functionality of porous metals derives from the combinations of these two distinct materials and, in essence, their specific porous structures. Porous metals are produced by a variety of techniques, including foaming, casting, and powder metallurgy. Recent advances in additive manufacturing have also added impetus to the field. Porous metals are finding new applications in many sectors, such as the aerospace, automotive, construction, and energy industries, for their unique properties. The applications of porous metals can largely be categorized into two classes—mechanical energy absorption due to their ability to undergo a significant amount of plastic deformation, and non-mechanical functions such as filtering, heat dissipation, electromagnetic shielding, sound absorption, etc. The functional applications are particularly sensitive to the nature of the porous structure, characterized by a variety of parameters, such as pore size, porosity, surface area, and fluid permeability.

This reprint is a collection of 10 papers from a Special Issue on porous metals in the journal *Materials*. The original aim of the Special Issue was to introduce the most recent advances in the field of porous metals, with special emphasis on new fabrication methods, novel structures, new properties, and new applications. These papers involve several manufacturing technologies, including 3D printing, melt infiltration, and powder sintering. They cover a range of different materials, including titanium, aluminum, copper, magnesium alloys, composites, and sandwich structures. They investigate a range of properties that are either uniquely derived from porous structures or heavily influenced by porous structures, including mechanical properties, energy absorption in explosions, friction, the capillary effect, the heat transfer coefficient, and catalytic efficiency. This is not intended to provide an exhaustive list of topics covered in the reprint but to bring to the attention of readers the diverse and comprehensive nature of this collection. We are delighted that the original aim was fully met, and we hope our readers will find the research contained in this reprint informative and useful.

We would like to thank Ms. Jean Yan, the Section Managing Editor, for her hard work behind the scenes. This reprint would not have been published without her assistance.

Yuyuan Zhao and Huiping Tang

Guest Editors

materials

MDPI

Communication

Effortless Fabrication of Nanofused HKUST-1 for Enhanced Catalytic Efficiency in the Cyanosilylation of Aldehyd

Tian Zhao

School of Chemical Engineering and Technology, Sun Yat-sen University, Zhuhai 519082, China; zhaot33@mail.sysu.edu.cn

Abstract: HKUST-1 (HKUST = Hong Kong University of Science and Technology) is one of the most recognized metal-organic frameworks (MOFs) based on copper and trimesate, extensively studied for a variety of applications, such as gas storage, separation, adsorption, electrocatalysis, drug delivery, sensor and photodegradation, etc. In this work, we introduce a novel nanofused HKUST-1, referred to as N-CuBTC (BTC = trimesate), which has been synthesized with the hydrothermal method at room temperature (typical synthesis temperature is from 80~120 °C). The resulting N-CuBTC features an irregular particle morphology, with numerous crystals clustering together and edges that have fused, creating a hierarchical pore structure. In contrast to the traditional micro-sized octahedral HKUST-1 (named as M-CuBTC), N-CuBTC displays a unique clumped morphology, where the HKUST-1 crystals are seamlessly integrated into a cohesive structure. This innovative formation significantly enhances mass transfer capabilities and porosity accessibility. Consequently, N-CuBTC demonstrates markedly improved catalytic performance in the cyanosilylation of aldehydes.

Keywords: metal-organic frameworks; HKUST-1; nanofused structure; cyanosilylation of aldehydes; enhanced catalytic efficiency

check for updates

Academic Editor: Lucjan Chmielarz

Received: 8 February 2025

Revised: 21 February 2025

Accepted: 24 February 2025

Published: 2 March 2025

Citation: Zhao, T. Effortless Fabrication of Nanofused HKUST-1 for Enhanced Catalytic Efficiency in the Cyanosilylation of Aldehyd. *Materials* **2025**, *18*, 1131. https://doi.org/10.3390/ma18051131

1. Introduction

Metal-organic frameworks (MOFs) represent a class of porous materials, uniquely formed from the combination of inorganic metal ions with organic ligands [1,2]. These materials are characterized by their remarkably high specific surface areas, customizable pore structures, and the presence of unsaturated metal sites [3,4]. Such features bestow them with outstanding capabilities across a variety of applications, including catalysis, drug delivery, adsorption, and energy storage and conversion [5,6].

HKUST-1, in particular, is a notable three-dimensional porous structure composed of copper ions and trimesate ligands, represented by the empirical formula 3D-$[Cu_3(btc)_2(H_2O)_3]$, where btc refers to trimesate [7,8]. Due to its straightforward synthesis process, high porosity, and impressive specific surface area (exceeding 1000 $m^2\ g^{-1}$), along with its stability and resistance to water, HKUST-1 has emerged as one of the most extensively researched MOFs [9,10]. Its exceptional properties make it an ideal candidate for a range of applications, including gas storage, gas separation, catalysis, drug delivery, and various sensing technologies [11,12]. Recent studies have centered on the development and utilization of hierarchically porous HKUST-1, also known as HP-HKUST-1 [13–15]. The techniques employed to fabricate these hierarchical structures, which primarily include etching [16,17], modulation [18,19], and mixed-linker strategies [20], frequently lead to the creation of mesopores and macropores. However, these larger pores can be interpreted as defects within the otherwise uniform micron-sized HKUST-1 crystals. As a result, the HP-HKUST-1 obtained through these methods often exhibits reduced structural integrity, primarily manifesting as a decrease in crystallinity and a lower specific surface area [21,22]. Thus, there is an evident need for a simpler, more efficient, and economically viable template-free approach to synthesize fully connected HP-HKUST-1.

This research introduces an innovative technique for the fabrication of a fully nanofused structural variant of HKUST-1, which is referred to as N-CuBTC. This method exploits the alteration of the copper source used in the synthesis. Remarkably, under optimal conditions, using copper acetate as Cu source at ambient temperature and stirring for 24 h, the crystals of HKUST-1 can be reduced to the nanoscale while achieving complete fusion, leading to a nanofused structure characterized by entirely interconnected pore channels, known as N-CuBTC. When compared to the original micron-sized HKUST-1 (referred to as M-HKUST-1), the resulting hierarchical pore structure in N-CuBTC exhibited remarkable accessibility. This enhanced accessibility plays a significant role in improving its catalytic efficiency, particularly in the cyanosilylation reactions of aldehydes.

2. Materials and Methods

2.1. Raw Materials and Reagents

$Cu(NO_3)_2{\cdot}3H_2O$ (99.5%, AR), $Cu(CH_3COO)_2{\cdot}H_2O$ (99.5%, AR), trimesic acid ($C_9H_6O_6$, 99.5%, AR), ethanol (C_2H_5OH, 99.5%, AR), benzaldehyde (C_7H_6O, 99+%, AR), trimethylsilylcyanide (Me_3SiCN, 97%, AR), heptane (C_7H_{16}, 99%, AR), and hexadecane ($C_{16}H_{34}$, 99%, AR) were utilized in this study. Water with a residual conductivity of 0.1 μS/cm was obtained using a GRO-10L water purification system. All chemicals were sourced from Shanghai Aladdin Biochemical Technology Co., Ltd. (Shanghai, China) and were used as received without additional purification.

2.2. Sample Synthesis and Purification

Synthesis of conventional octahedral HKUST-1 (M-CuBTC): In this procedure, 1.21 g of copper nitrate trihydrate (equivalent to 5 mmol) and 0.63 g of trimesic acid (which corresponds to 3 mmol) were each individually dissolved in 15 mL of distilled water. Following this, the two solutions were combined and stirred (1000 r/min) continuously for a duration of 24 h at room temperature (25 °C). After thorough mixing, the resulting blue crystals were isolated using centrifugation. These crystals were then washed with 30 mL of anhydrous ethanol through ultrasonic agitation for 1 h, with the ethanol being replaced every 30 min. Once the washing procedure was completed, the product was again collected via centrifugation and subsequently placed in a vacuum drying oven set to 80 °C to dry overnight.

Synthesis of nanofused HKUST-1 (N-CuBTC): For the preparation of N-CuBTC, 1.00 g of copper acetate monohydrate (5 mmol) was dissolved in 15 mL of deionized water, while 0.63 g of trimesic acid (3 mmol) was dissolved separately in another 15 mL of deionized water. The two prepared solutions were then mixed and stirred (1000 r/min) for a continuous 24 h period at room temperature (25 °C). After stirring was complete, the blue product was collected by centrifugation. The purification of N-CuBTC was carried out following the same protocol utilized for M-CuBTC.

Cyanosilylation of benzaldehyde

A total of 425 mg (4 mmol) of benzaldehyde, along with 794 mg (8 mmol) of trimethylsilylcyanide (TMSCN), and 10 mg of the MILs catalyst were mixed in 15 mL of heptane, with 1 mL of hexadecane serving as the internal standard [23]. The reaction mixture was stirred at a temperature of 313 K, and its progress was periodically assessed using an Agilent Technologies 7890A gas chromatography (GC) system.

2.3. Detection and Characterization

2.3.1. X-Ray Diffraction (XRD)

The samples were analyzed using a Bruker D8 Advance X-ray diffractometer (Bruker, Karlsruhe, Germany) equipped with a Cu target and Kα radiation (λ = 1.54182 nm), operating at a voltage of 30 kV. The measurement was conducted over a 2 h duration within a 5° to 80° test angle range. The XRD was used for phase identification of the samples.

2.3.2. Surface Area and Pore Volume Measurement

The specific surface area and pore volume of the samples were determined using a specific surface area analyzer (NOVA-4200e, Quantachrome, Boynton Beach, FL, USA). Before the analysis, all samples underwent a standardized pretreatment, which included drying in a vacuum at 120 °C for 2 h. The chosen temperature of 120 °C and the duration of 2 h for the vacuum drying pretreatment were specifically selected to effectively remove the crystallization water from the samples. This temperature is sufficiently high to promote the evaporation of water without compromising the structural integrity of the material. The 2 h duration ensures thorough dehydration while preventing potential thermal decomposition or alteration of the sample's properties. The measurements were carried out within a pressure range of 10.1325 to 101.325 kPa, employing high-purity nitrogen (N_2) as the adsorbent at a temperature of 120 °C.

2.3.3. Transmission Electron Microscopy (TEM)

The particle size and microscopic morphology of the samples were investigated using transmission electron microscopy (TEM, Talos F200X, FEI, Hillsboro, OR, USA) with an FEI Talos F200X instrument (Thermo Fisher Scientific, Waltham, MA, USA). For the examination, samples were dispersed in ethanol and then deposited onto a standard copper grid.

2.3.4. Scanning Electron Microscopy (SEM)

To assess the microscopic morphology and particle dimensions, we employed a Zeiss Gemini 300 scanning electron microscope (SEM, Gemini 300, Zeiss, Jena, Germany). Prior to the imaging process, samples were prepared by placing them on a conductive gel substrate and coating them with a layer of gold to enhance conductivity.

2.3.5. Fourier Transform Infrared Spectroscopy (FTIR)

Infrared spectra of the samples were obtained using an infrared spectrometer (Vertex 80 V, Bruker, Karlsruhe, Germany) over a wavenumber range spanning from 400 to 4000 cm^{-1}. The resolution during the FTIR scanning was set at 1 cm^{-1}.

2.3.6. Gas Chromatography (GC)

The products resulting from the catalytic reaction were identified and analyzed utilizing an Agilent Technologies 7890A gas chromatography (GC, Agilent, Palo Alto, CA, USA) system.

3. Results and Discussion

The morphologies and structural characteristics of M-CuBTC and N-CuBTC were investigated through electron microscopic imaging, as illustrated in Figure 1. The M-CuBTC, serving as the control sample, displayed the quintessential octahedral crystalline form on a micron scale (refer to Figure 1A,C), featuring a consistent crystal size of approximately 12 μm across the sample.

Figure 1. SEM images of (**A**) M-CuBTC and (**B**) N-CuBTC. TEM images of (**C**) M-CuBTC and (**D**) N-CuBTC.

In contrast, the N-CuBTC exhibited dramatically different features, showcasing aggregates of irregularly shaped nanocrystals, each measuring approximately 50 nm. These nanocrystals displayed significant evidence of intergrowth and fusion (see Figure 1B,D). It was observed that approximately ~100% of the nanocrystals were fully fused together (Figure 1B,D). Additionally, surface area measurements revealed that the specific surface area of N-CuBTC was only decreased by approximately 70 m^2/g compared to the non-fused counterparts (M-CuBTC, Table 1). This unique arrangement enhances the material's potential for various applications due to its improved accessibility and porosity.

Table 1. Pore structure information for M-CuBTC and N-CuBTC.

Sample	Yield /(%) [a]	Particle Size /(nm) [b]	S_{BET} /(m^2 g^{-1}) [c]	$S_{Langmuir}$ /(m^2 g^{-1})	V_{pore} /(cm^3 g^{-1}) [d]
M-CuBTC	90.2	~12000	1290	3150	0.47
N-CuBTC	93.1	~50	1220	2990	0.58

[a] The yield is calculated based on a copper to BTC ratio of 1.67:1. [b] The average particle size was assessed using Gaussian statistical methods, with the number in parentheses indicating the standard deviation. [c] The surface area was determined from N_2 adsorption isotherms at 77 K within the pressure range of $0.05 < P/P_0 < 0.2$, with an estimated standard deviation of ± 50 m^2 g^{-1}. [d] The pore volume was derived from N_2 adsorption isotherms at 77 K, specifically at $P/P_0 = 0.95$, for pores measuring ≤ 20 nm.

It is clear that the samples obtained from different copper precursors exhibit notable differences in their morphology and structural characteristics. It has been established in prior studies that the type of copper source employed plays a crucial role in determining the nucleation dynamics and size distribution of the resulting metal-organic framework (MOF), HKUST-1. Specifically, the utilization of copper nitrate has been associated with the formation of larger HKUST-1 particles due to its slower nucleation rate. In contrast, the employment of copper acetate leads to the generation of nano-sized HKUST-1 particles, as the accelerated nucleation process facilitates a rapid formation of numerous smaller particles [24,25].

In the current investigation, it is proposed that the nanofusion mechanism contributing to the formation of N-CuBTC nanocrystals adheres to this established pattern. When copper acetate is used as the precursor, a swift generation of nanocrystallites occurs, which significantly increases the likelihood of nanofusion owing to the heightened surface energy of these small particles. The elevated surface energy results in stronger particle interactions, thereby promoting their fusion into larger aggregates [24,25]. Moreover, it is reported that the application of rapid stirring during the synthesis process can further facilitate the fusion of MOF nanoparticles [26]. This dynamic mixing likely enhances the opportunities for particle collisions and fusion, thereby contributing to the final characteristics of the synthesized N-CuBTC nanocrystals. Conversely, in reactions involving copper nitrate, the gradual nucleation rate leads to a different particle morphology observed in the M-CuBTC samples, where larger aggregates are formed without the same level of rapid surface interactions. In conclusion, the combined effects of precursor selection, nucleation dynamics, and external factors such as stirring are crucial in dictating the morphology and size of the synthesized nanocrystals, confirming that similar mechanisms are operational in the formation of N-CuBTC.

The X-ray diffraction (XRD) analysis of the samples provided further insights, revealing that both M-CuBTC and N-CuBTC closely corresponded to the simulated XRD pattern of HKUST-1. This confirmed the purity of the phase for all samples studied (refer to Figure 2A). Notably, as we transitioned from M-CuBTC to N-CuBTC, a discernible decrease in the peak intensities of the XRD patterns was observed, along with a gradual softening of peak shapes (see Figure 2A). This trend suggests a reduction in particle size, which aligns well with the observations made through electron microscopic imaging, reinforcing the contrast between the two materials and their respective structural characteristics.

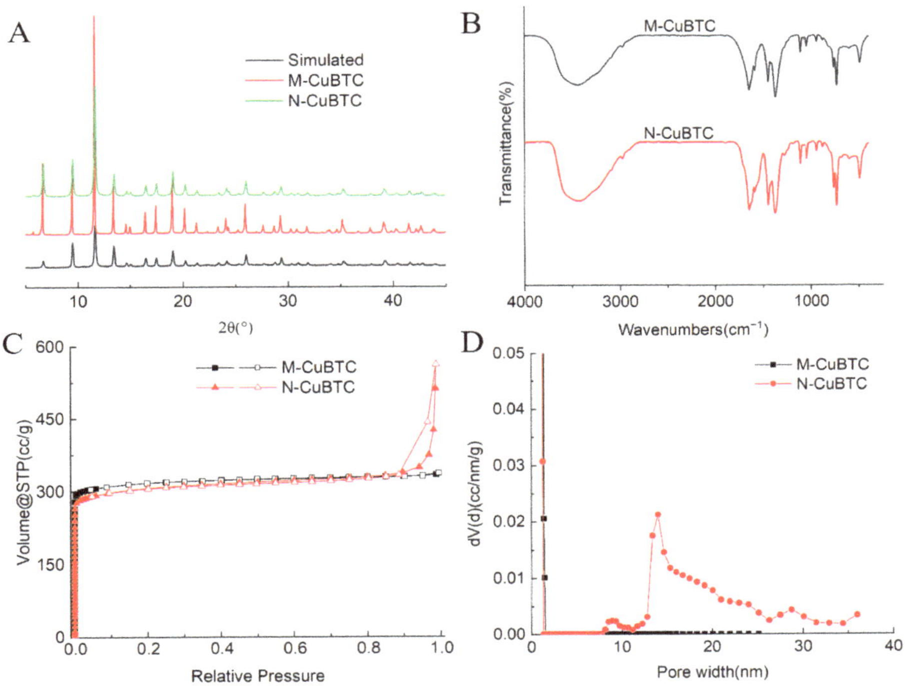

Figure 2. (**A**) XRD patterns and (**B**) FTIR spectra of M-CuBTC and N-CuBTC. (**C**) N_2 adsorption–desorption curves and (**D**) NLDFT pore size distribution (PSD) curves for M-CuBTC and N-CuBTC.

The application of Fourier transform infrared spectroscopy (FTIR) allowed us to investigate the coordination of trimesate with Cu^{2+} ions in the HKUST-1 framework, yielding results that align well with the early literature [27] (Figure 2B). The absorption peak observed at 730 cm^{-1} does indeed correspond to the presence of Cu ion clusters within the crystal structure, and it aligns well with the reference spectra reported in the literature. This consistency supports the accuracy of the peak assignments and further validates the findings regarding the structural characteristics of the material [28]. Furthermore, the spectra exhibited absorption peaks in the range of 1300 to 1700 cm^{-1}, indicative of C-O vibrational modes associated with the carboxylic functional groups. Specifically, the peak at 1451 cm^{-1} was linked to the symmetric stretching vibration of the O=C-O bond, while the peak at 1634 cm^{-1} corresponded to the asymmetric stretching of the same bond.

Additionally, the peaks at 1374 cm^{-1} and 1573 cm^{-1} represented vibrational modes of C=C, signifying the successful intercalation of the BTC ligand within the HKUST-1 structure. A broad absorption peak situated at 3421 cm^{-1} indicated H-O stretching vibrations, which suggests a considerable presence of adsorbed water molecules within the HKUST-1 matrix. These spectral data imply that the coordination of the H_3btc ligand with Cu^{2+} occurs in a bidentate fashion, resulting in the formation of the $[Cu_2(COO)_4]$ structural unit, commonly referred to as the pulley unit. This structural motif has the potential to be activated through heating or under vacuum, thereby generating a substantial number of unsaturated coordination sites of Cu^{2+}, which could function as catalytic Lewis acid sites.

The nitrogen adsorption–desorption isotherms for both M-CuBTC and N-CuBTC are presented in Figure 2C, with the accompanying porosity parameters provided in Table 1. The nitrogen adsorption–desorption isotherms for both M-CuBTC and N-CuBTC are indeed classified as typical Type I (a) isotherms. This classification indicates the presence of microporosity within the materials, highlighting their ability to adsorb nitrogen at low pressures [29]. The adsorption and desorption curves for M-CuBTC closely overlapping, demonstrating minimal hysteresis consistent with the characteristics of microporous materials. In contrast, the isotherms for N-CuBTC displayed noticeable hysteresis loops between the adsorption and desorption phases in the P/P_0 range of 0.9 to 1.0, a behavior typical of meso- and macropores exist in the material [29] (refer to Figure 2C).

The pore size distribution (PSD) analyses for both M-CuBTC and N-CuBTC, calculated using nonlocal density functional theory (NLDFT) with a "slit-pore model," are illustrated in Figure 2D. It is noteworthy that N-CuBTC contains not only its intrinsic micropores but also additional mesopores and macropores, confirming the observations made through electron microscopy. Conversely, M-CuBTC comprises exclusively micropores. This distinct difference suggests that N-CuBTC may offer considerable advantages in applications requiring swift mass transport, such as various catalytic processes.

The presence of coordinated water molecules within the HKUST-1 structure can be effectively removed when subjected to vacuum conditions, resulting in the formation of active Lewis acid sites [30,31]. This characteristic renders HKUST-1 a valuable catalyst for a variety of chemical reactions, including the cyanosilylation process involving aldehydes [32].

To evaluate the catalytic performance of the synthesized materials M-CuBTC and N-CuBTC, reactions were conducted based on this methodology, with the reaction scheme illustrated in Scheme 1A. The results regarding the catalytic efficiency of both M-CuBTC and N-CuBTC are compiled in Table 2, while the substrate conversion over time is depicted in Figure 3A. Specifically, during the cyanosilylation of benzaldehyde, N-CuBTC achieved a remarkable conversion rate of 99.4%, which indicates a substantial 20% improvement over the performance of M-CuBTC as shown in Figure 3A and summarized in Table 2. Thus, N-CuBTC displayed considerably enhanced catalytic activity compared to M-CuBTC. Also, compare to other work, N-CuBTC exhibited quite good efficient and reusability (details please see Table 2).

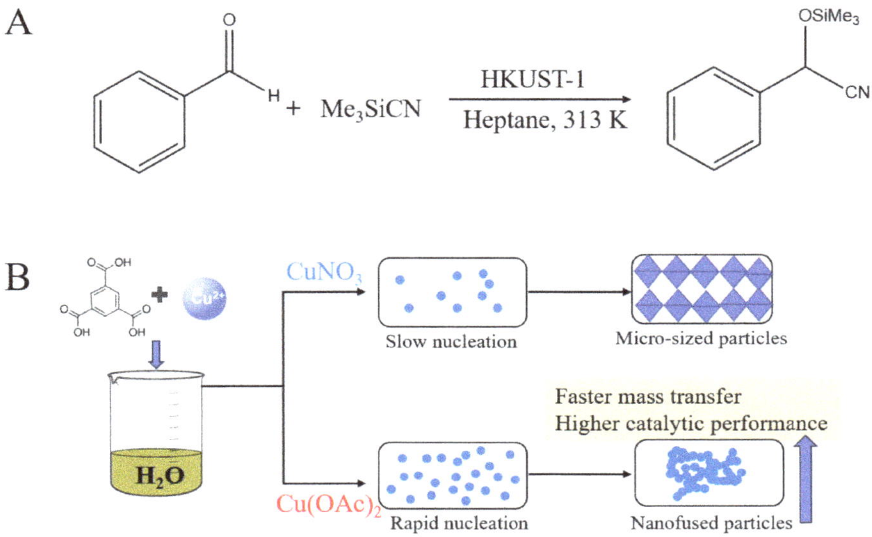

Scheme 1. (**A**) The schemes of cyanosilylation of aldehyde reaction. (**B**) The synthesis procedure and possible mechanism for M-CuBTC and N-CuBTC.

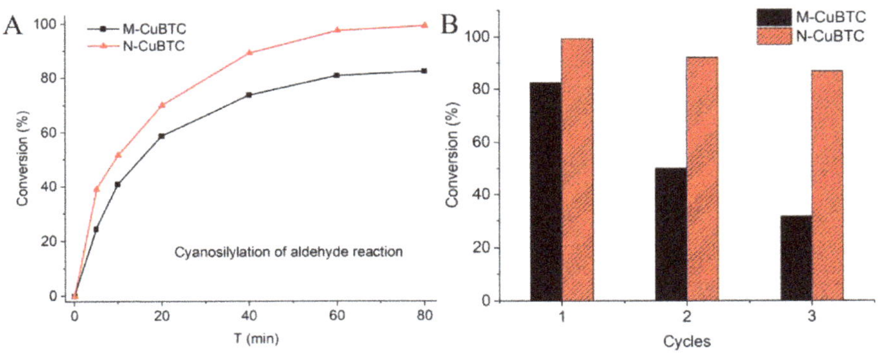

Figure 3. (**A**) Time-dependent conversion of benzaldehyde using M-CuBTC and N-CuBTC. (**B**) Comparison of conversions achieved with M-CuBTC and N-CuBTC over three reaction cycles.

Furthermore, N-CuBTC displayed impressive reusability in catalytic applications; after undergoing three reaction cycles in the cyanosilylation of benzaldehyde, it maintained a conversion rate exceeding 85% (see Figure 3B). In contrast, the performance of M-CuBTC as a catalyst deteriorated significantly after three cycles, with conversion dropping to below 40% (as illustrated in Figure 3B).

These findings clearly indicate that the distinctive nanofused hierarchical structure of N-CuBTC plays a critical role in enhancing its catalytic performance. The fully interconnected hierarchical pores increase the accessibility of the active sites located within the structure, facilitating improved mass transfer rates for the reacting substrate [33,34], which in turn significantly boosts the overall catalytic efficiency (refer to Scheme 1B) [35].

Table 2. Summary on catalytic performances for M-CuBTC/N-CuBTC and the comparison with other work.

Run	Catalysit	Reaction	Advantages & Disadvantages	Conversion (%)	Reference
1	M-CuBTC	Cyanosilylation of aldehyde reaction	highly efficient and reusability	82.5	This work
2	N-CuBTC			99.4	
3	$[Zn(H_2L_2)]_2$		efficient, versatile/ long time	97	[36]
4	3-AC		efficient and reusability/surface modification limitations	94.3	[37]
5	$[Co(H_2L_1)]_2$		efficient, non-solvent/ moderate yield	92	[38]
6	MnI/DAPTA		efficient, green solvent/ metal dependency	99	[39]
7	Co complexe		versatile/low efficient	77.4	[40]
8	F-MIL		efficient/limited-reusability	98.4	[41]
9	L2 + Ph_3PO		high enantioselectivity/ challenging optimization	71	[42]
10	TBD		efficient, low-catalyst-loading/	98	[43]
11	PF6		efficient/limited-to-aldehydes	99	[44]
12	$Cu(ClO_4)_2{\cdot}6H_2O$		efficient/electronic-effect-sensitive	98	[45]

4. Conclusions

In conclusion, this research successfully illustrates the synthesis of an innovative nanofused variant of HKUST-1, designated as N-CuBTC. This achievement was made possible by altering the copper precursor during the hydrothermal synthesis process. The resulting N-CuBTC displayed a sophisticated hierarchical porous architecture, which markedly improved its catalytic efficacy in the cyanosilylation of aldehydes. The catalyst attained an impressive conversion rate of 99.4% along with a turnover frequency (TOF) reaching 4.97 mmol g^{-1} min^{-1}, demonstrating its high efficiency.

Moreover, N-CuBTC exhibited outstanding reusability, maintaining its catalytic performance across several reaction cycles, underscoring its practical potential for sustainable applications. Looking ahead, future studies could focus on further refining the synthesis techniques and exploring the applicability of the hierarchical porous catalyst in various chemical transformation processes. Specifically, transformations such as the selective oxidation of alcohols to aldehydes and ketones could benefit from the increased accessibility provided by the hierarchical pore structures. Additionally, the catalyst may be advantageous in the catalytic conversion of biomass-derived compounds, which often require efficient mass transfer due to the larger size of substrates. Furthermore, applications in coupling reactions, such as Suzuki or Heck reactions, could also leverage the unique properties of hierarchical pores to enhance reaction rates. These potential applications highlight the versatility of the catalyst and its promising role in advancing catalytic efficiency across different reaction types.

Funding: The work was supported by the National Natural Science Foundation of China (51802094).

Institutional Review Board Statement: Not applicable.

Informed Consent Statement: Not applicable.

Data Availability Statement: The original contributions presented in this study are included in the article. Further inquiries can be directed to the corresponding author.

Conflicts of Interest: The authors declare that they have no known competing financial interests or personal relationships that could have appeared to influence the work reported in this paper.

References

1. Zhao, T.; Xiao, P.; Nie, S.; Luo, M.; Zou, M.; Chen, Y. Recent progress of metal-organic frameworks based high performance batteries separators: A review. *Coord. Chem. Rev.* **2024**, *502*, 215592. [CrossRef]
2. Zhao, T.; Zou, M.; Xiao, P.; Luo, M.; Nie, S. Template-Free Synthesis and Multifunctional Application of Foam HKUST-1. *Inorg. Chem.* **2023**, *62*, 14659–14667. [CrossRef] [PubMed]
3. Zhao, T.; Nie, S.; Luo, M.; Xiao, P.; Zou, M.; Chen, Y. Research progress in structural regulation and applications of HKUST-1 and HKUST-1 based materials. *J. Alloys Compd.* **2024**, *974*, 172897. [CrossRef]
4. Zou, M.; Zhu, H.; Dong, M.; Zhao, T. Template Method for Synthesizing Hierarchically Porous MIL-101(Cr) for Efficient Removal of Large Molecular Dye. *Materials* **2022**, *15*, 5763. [CrossRef]
5. Doan, H.V.; Amer Hamzah, H.; Karikkethu Prabhakaran, P.; Petrillo, C.; Ting, V.P. Hierarchical Metal-Organic Frameworks with Macroporosity: Synthesis, Achievements, and Challenges. *Nanomicro Lett.* **2019**, *11*, 54. [CrossRef]
6. Cai, G.; Yan, P.; Zhang, L.; Zhou, H.C.; Jiang, H.L. Metal-Organic Framework-Based Hierarchically Porous Materials: Synthesis and Applications. *Chem. Rev.* **2021**, *121*, 12278–12326. [CrossRef]
7. Raffah, B.M.; Hassan, H.; Iqbal, M.W.; Al-Hadeethi, Y. Triumph of trifunctionality: Siloxene@HKUST-1-PANI unleashing multifunctionality in batteries, supercapacitors, and hydrogen evolution. *Inorg. Chem. Commun.* **2025**, *171*, 113514. [CrossRef]
8. Loera-Serna, S.; Cortes-Suarez, J.; Sanchez-Salas, R.; Ramirez-Rosales, D.; Oliver-Tolentino, M.; Ramos-Fernandez, E.V. CO_2 adsorption on a water-resist HKUST-1 by incorporation of Graphene Oxide. *Adsorpt.-J. Int. Adsorpt. Soc.* **2025**, *31*, 1–22. [CrossRef]
9. Mohammadnejad, M.; Moeinipour, A. Study on Efficiency of HKUST-1 Metal-organic Framework and Modified HKUST- 1s for Removal of Meloxicam from Aqueous Solution. *Anal. Bioanal. Chem. Res.* **2024**, *11*, 77–90.
10. Wang, B.; Ke, J.; Zhang, J. Thermomechanical properties of metal-organic framework HKUST-1 crystals. *J. Mater. Chem. A* **2024**, *12*, 15071–15081. [CrossRef]
11. Ran, Q.; Tao, B.; Li, M.; Zheng, K.; She, Y.; Wu, W.; Li, Z.; Luo, D.; Xu, X. Fundamentals of metal-organic framework-based photocatalysts: The role of chemical diversity and structure engineering. *Coord. Chem. Rev.* **2025**, *525*, 216324. [CrossRef]
12. Shariatinia, Z.; Karimzadeh, Z. Recent advances on oxidation of alcohols over various types of materials as effective catalysts: An overview. *Coord. Chem. Rev.* **2025**, *526*, 216372. [CrossRef]
13. Hara, Y.; Sakaushi, K. Emergent electrochemical functions and future opportunities of hierarchically constructed metal-organic frameworks and covalent organic frameworks. *Nanoscale* **2021**, *13*, 6341–6356. [CrossRef] [PubMed]
14. Zhao, T.; Yu, J.; Xiao, P.; Nie, S.; Peng, S.; Chen, J.; Luo, F.; Janiak, C.; Chen, Y. Preparation of Hierarchical Porous ZIF-67 and Its Application in Zinc Battery Separator. *Chemistry* **2024**, *6*, 1363–1373. [CrossRef]
15. Qiu, X.; Wang, R. From construction strategies to applications: Multifunctional defective metal-organic frameworks. *Coord. Chem. Rev.* **2025**, *526*, 216356. [CrossRef]
16. Yuan, S.; Tu, Y.; Yu, R.; Nie, F. Tunable Chemiluminescence Kinetics with Hierarchically Structured HKUST-1 and Its Sensing Application for Concanavalin A Analysis. *ACS Appl. Mater. Interfaces* **2024**, *16*, 63760–63768. [CrossRef]
17. Kubo, M.; Kitano, T.; Shimada, M. Spray synthesis of hierarchical porous metal-organic framework HKUST-1 with soft templates and methylene blue adsorption performance. *Adv. Powder Technol.* **2024**, *35*, 104280. [CrossRef]
18. Wang, N.; Wei, Y.; Chang, M.; Liu, J.; Wang, J.-X. Macro-Meso-Microporous Metal-Organic Frameworks: Template-Assisted Spray Drying Synthesis and Enhanced Catalysis. *ACS Appl. Mater. Interfaces* **2022**, *14*, 10712–10720. [CrossRef]
19. Bai, K.; Fan, S.; Chen, Y.; Wang, Y.; Chen, J.; Mai, Z.; Liu, J.; Deng, L.; Xiao, Z. Membrane adsorber with hierarchically porous HKUST-1 immobilized in membrane pores by flowing synthesis. *J. Membr. Sci.* **2022**, *650*, 120424. [CrossRef]
20. Liu, T.; Liu, Y.; Yao, L.; Yang, W.; Tian, L.; Liu, H.; Liu, D.; Wang, C. Controllable formation of meso- and macropores within metal-organic framework crystals via a citric acid modulator. *Nanoscale* **2018**, *10*, 13194–13201. [CrossRef]
21. Xiao, P.; Nie, S.; Luo, M.; Chen, J.; Luo, F.; Zhao, T.; Liu, Y.-J. Application of Metal-Organic Frameworks for Battery Separators. *Prog. Chem.* **2024**, *36*, 1217–1236.
22. Zhao, T.; Li, S.; Xiao, Y.-X.; Janiak, C.; Chang, G.; Tian, G.; Yang, X.-Y. Template-free synthesis to micro-meso-macroporous hierarchy in nanostructured MIL-101(Cr) with enhanced catalytic activity. *Sci. China-Mater.* **2021**, *64*, 252–258. [CrossRef]
23. Zhang, Z.; Chen, J.; Bao, Z.; Chang, G.; Xing, H.; Ren, Q. Insight into the catalytic properties and applications of metal–organic frameworks in the cyanosilylation of aldehydes. *RSC Adv.* **2015**, *5*, 79355–79360. [CrossRef]
24. Xin, C.; Zhan, H.; Huang, X.; Li, H.; Zhao, N.; Xiao, F.; Wei, W.; Sun, Y. Effect of various alkaline agents on the size and morphology of nano-sized HKUST-1 for CO_2 adsorption. *RSC Adv.* **2015**, *5*, 27901–27911. [CrossRef]
25. Zhao, T.; Dong, M.; Zhao, Y.; Liu, Y.J. Preparation and Application of Nano-Sized Metal-Organic Frameworks. *Prog. Chem.* **2017**, *29*, 1252–1259.
26. Yue, Y.; Qiao, Z.A.; Fulvio, P.F.; Binder, A.J.; Tian, C.; Chen, J.; Nelson, K.M.; Zhu, X.; Dai, S. Template-free synthesis of hierarchical porous metal-organic frameworks. *J. Am. Chem. Soc.* **2013**, *135*, 9572–9575. [CrossRef]
27. Kangra, B.; Jatrana, A.; Maan, S.; Chauhan, S.; Kumar, V. Effective adsorption of chlorpyrifos pesticides by HKUST-1 metal-organic framework. *J. Chem. Sci.* **2022**, *134*, 104. [CrossRef]

28. Garg, D.; Rekhi, H.; Kaur, H.; Singh, K.; Malik, A.K. A Novel Method for the Synthesis of MOF-199 for Sensing and Photocatalytic Applications. *J. Fluoresc.* **2022**, *32*, 1171–1188. [CrossRef]
29. Thommes, M.; Kaneko, K.; Neimark, A.V.; Olivier, J.P.; Rodriguez-Reinoso, F.; Rouquerol, J.; Sing, K.S.W. Physisorption of gases, with special reference to the evaluation of surface area and pore size distribution (IUPAC Technical Report). *Pure Appl. Chem.* **2015**, *87*, 1051–1069. [CrossRef]
30. Guo, L.; Du, J.; Li, C.; He, G.; Xiao, Y. Facile synthesis of hierarchical micro-mesoporous HKUST-1 by a mixed-linker defect strategy for enhanced adsorptive removal of benzothiophene from fuel. *Fuel* **2021**, *300*, 120955. [CrossRef]
31. Yang, L.; Wang, K.; Guo, L.; Hu, X.; Zhou, M. Unveiling the potential of HKUST-1: Synthesis, activation, advantages and biomedical applications. *J. Mater. Chem. B* **2024**, *12*, 2670–2690. [CrossRef] [PubMed]
32. Larrea, E.S.; Fernandez de Luis, R.; Orive, J.; Iglesias, M.; Arriortua, M.I. [NaCu(2,4-HPdc)(2,4-Pdc)] Mixed Metal-Organic Framework as a Heterogeneous Catalyst. *Eur. J. Inorg. Chem.* **2015**, *28*, 4699–4707. [CrossRef]
33. Zhao, T.; Nie, S.; Xiao, P.; Peng, S.; Chen, J.; Yu, J.; Luo, F.; Chen, Y. Exploring the influence of MIL-101(Cr) morphologies on the efficacy of cellulose separators for zinc ion battery performance. *J. Membr. Sci.* **2025**, *720*, 123788. [CrossRef]
34. Feng, L.; Wang, K.-Y.; Powell, J.; Zhou, H.-C. Controllable Synthesis of Metal-Organic Frameworks and Their Hierarchical Assemblies. *Matter* **2019**, *1*, 801–824. [CrossRef]
35. Zou, M.; Dong, M.; Luo, M.; Zhu, H.; Zhao, T. Nanofused Hierarchically Porous MIL-101(Cr) for Enhanced Methyl Orange Removal and Improved Catalytic Activity. *Materials* **2022**, *15*, 3645. [CrossRef]
36. Estifanos Filkale, A.; Kumar Gangwar, M.; Kumar Pandey, I.; Butcher, R.J. Biphenol-based tetradentate N2O2 donor scaffold binuclear Zn(II) and trinuclear Cu(II) Complexes: Synthesis, Structure, Characterizations, and catalytic cyanosilylation of aldehydes at room temperature. *Inorg. Chim. Acta* **2024**, *559*, 121801.
37. Mahmoud, A.G.; Librando, I.L.; Paul, A.; Carabineiro, S.A.C.; Ferraria, A.M.; Botelho do Rego, A.M.; Guedes da Silva, M.F.C.; Geraldes, C.F.G.C.; Pombeiro, A.J.L. Novel organotin-PTA complexes supported on mesoporous carbon materials as recyclable catalysts for solvent-free cyanosilylation of aldehydes. *Catal. Today* **2023**, *423*, 114270. [CrossRef]
38. Filkale, A.E.; Gangwar, M.K.; Butcher, R. Well-Defined Binuclear Co(II) Complexes: Synthesis, Structure, Characterizations and Catalytic Applications in Cyanosilylation of Various Aldehydes under Solvent-Free Conditions. *Asian J. Org. Chem.* **2023**, *12*, e202300085.
39. Mahmoud, A.G.; da Silva, M.F.C.G.; Pombeiro, A.J.L. Organomanganese/amido-phosphine (DAPTA) catalyst for rapid cyanosilylation of aldehydes in glycerol and solvent-free conditions at room temperature. *Catal. Today* **2023**, *418*, 114056.
40. Gurbanov, A.V.; Mahmoudi, G.; Zangrando, E.; Zubkov, F.I.; Hasanov, K.I.; Mahmudov, K.T. Bridging linker installation strategy in the engineering of copper(ii) complexes: Activation of nitriles, efficient catalysts for the cyanosilylation of aldehydes. *CrystEngComm* **2024**, *26*, 6692–6700.
41. Zhao, T. Facile synthesis of foam MIL-101(Cr) with the enhanced catalytic capability. *Chem. Phys. Lett.* **2024**, *838*, 141089. [CrossRef]
42. Pawar, T.J.; Jimenez-Halla, J.O.C.; Martinez-Valencia, D.I.; Kokate, S.V.; Delgado-Alvarado, E.; Olivares-Romero, J.L. Investigation of Enantioselectivity Using TADDOL Derivatives as Chiral Ligands in Asymmetric Cyanation Reactions. *ACS Omega* **2024**, *9*, 29035–29040. [CrossRef]
43. Matsukawa, S.; Kimura, J. TBD-catalyzed cyanosilylation of aldehydes and ketones. *Synth. Commun.* **2016**, *46*, 1947–1952. [CrossRef]
44. Shen, Z.-L.; Ji, S.-J.; Loh, T.-P. Ionic liquid [omim][PF6] as an efficient and recyclable reaction media for the cyanosilylation of aldehydes without Lewis acid or any special activation. *Tetrahedron Lett.* **2005**, *46*, 3137–3139. [CrossRef]
45. Majhi, A.; Kim, S.S.; Kim, H.S. Copper perchlorate hexahydrate: A highly efficient catalyst for the cyanosilylation of aldehydes. *Appl. Organomet. Chem.* **2008**, *22*, 407–411. [CrossRef]

materials

MDPI

Article

Development of Tailored Porous Ti6Al4V Materials by Extrusion 3D Printing

Luis Olmos [1], Ana Silvia González-Pedraza [2,*], Héctor Javier Vergara-Hernández [2], Didier Bouvard [3], Monserrat Sofía López-Cornejo [2] and Rumualdo Servín-Castañeda [4]

1 Universidad Michoacana de San Nicolás de Hidalgo, INICIT, Fco. J. Mujica S/N, Morelia 58060, Michoacán, Mexico; luis.olmos@umich.mx

2 División de Estudios de Posgrado e Investigación, Tecnológico Nacional de México/I.T. Morelia, Av. Tecnológico #1500, Colonia Lomas de Santiaguito, Morelia 58120, Michoacán, Mexico; hector.vh@morelia.tecnm.mx (H.J.V.-H.); monserrat.lc@morelia.tecnm.mx (M.S.L.-C.)

3 University Grenoble Alpes, CNRS, SIMAP, 38000 Grenoble, France; didier.bouvard@grenoble-inp.fr

4 Universidad Autónoma de Coahuila, Facultad de Ingeniería Mecánica y Eléctrica Unidad Norte, Monclova 25720, Coahuila, Mexico; rumualdo.servin@uadec.edu.mx

* Correspondence: anna.silvia.1204@gmail.com

Abstract: Nowadays, metallic bone replacement is in high demand due to different issues, like sicknesses and accidents. Thus, bone implants are fabricated with tailored properties and microstructure for long-term use in the human body. To improve such implants, 3D printing is the most promising technique. Therefore, this work aims to evaluate the fabrication of porous materials by extrusion 3D printing of Ti6Al4V. Cylindrical samples were fabricated from pellets for metal injection molding of Ti6Al4V powders, creating hexagonal channels with three different sizes. The densification kinetics was evaluated by dilatometry tests, which enabled following the densification of the samples during the sintering cycle. Subsequently, the samples were characterized by scanning electron microscopy and X-ray computed tomography to analyze their microstructure. Compression tests evaluated the mechanical strength of sintered samples. It was found that the hexagonal shape during printing is better defined as the channel size increases. The results show similar behavior for each of the channel sizes during sintering; however, greater densification is obtained as the channel size decreases. Additionally, microporosity is obtained at the particle level, which is completely interconnected, ensuring the passage of fluids through the entire sample. On the other hand, as the channel size increases, Young's modulus and yield strength are considerably reduced. The main conclusion is that parts with two scales of porosity can be designed by the 3D printing extrusion process.

Keywords: additive manufacturing; porosity; computed tomography; permeability; sintering

Academic Editors: Yuyuan Zhao, Huiping Tang and Franz E. Weber

Received: 8 December 2024
Revised: 8 January 2025
Accepted: 13 January 2025
Published: 16 January 2025

Citation: Olmos, L.; González-Pedraza, A.S.; Vergara-Hernández, H.J.; Bouvard, D.; López-Cornejo, M.S.; Servín-Castañeda, R. Development of Tailored Porous Ti6Al4V Materials by Extrusion 3D Printing. *Materials* **2025**, *18*, 389. https://doi.org/10.3390/ma18020389

1. Introduction

In biomedical implants, porosity is a property that should not go unnoticed because it is through porosity that the diffusion of nutrients and vascularization occurs [1,2]. Besides improving the passage of fluids, porosity helps decrease mechanical properties to avoid the so-called "stress shielding." This is due to the high difference between the modulus of commonly used materials such as Ti6Al4V (Ti64) (90–115 GPa) [3], Co alloys (210–250 GPa) [4], and stainless steel (200 GPa) [5] compared to human bone (cortical bone 7–30 GPa [6]).

Different techniques have been developed for the fabrication of porous materials in order to reduce mechanical properties [7] and, at the same time, achieve better tissue–implant bonding [8]. To improve biomedical applications, porous materials classified as open cell are preferred because they show excellent interconnectivity and low mechanical resistance [9]. Among the techniques used to fabricate porous materials, the space holder technique is one of the most commonly used, in which particles with lower melting points are used as spacers. The pores' features depend on the shape, size, and quantity of space holders added. This technique's main advantage is the possibility of fabricating porous composites that can improve biocompatibility [10–12]. Karakurt et al. [13] fabricated a porous alloy (21–25% and 50–56%) of Ti-Nb-Zr to study the corrosion effect and in vitro tests. During the corrosion tests, they used Hank's solution, demonstrating better corrosion resistance for the alloys with low porosity (21–25%) in the same way they presented higher cell growth and adhesion. At the same time, the mechanical tests showed a low Young's modulus as the porosity percentage increased (50–56%). Another technique to consider for forming open-cell pores is through a liquid-state process, i.e., injecting gas (argon, air, nitrogen) directly into the molten metal and thus creating bubbles inside the metal and obtaining the porosity after cooling. Using this technique, the porosity can be controlled through the gas flow and the nozzle diameter, among others. However, controlling the pore size and morphology is impossible since only sub-spherical pores are obtained [14].

On the other hand, additive manufacturing (AM) has presented advantages over porous scaffold fabrication because it is possible to control the porosity while designing pores with complex architectures. Li et al. [15] fabricated Ti64 scaffolds by melting with pore sizes between 300 and 500 μm in four different morphologies. They demonstrated that as the channel size increases, the modulus is considerably reduced, as well as the role that channel morphology can play. On the other hand, they found that cell size and morphology play an important role in cell growth due to more cells for a size of 300 μm, with osseointegration and permeability being important factors in the fabrication of porous materials. Ouyang et al. [16] demonstrated how pore size influences the results of these properties as they fabricated titanium porous scaffolds by selective laser melting (SLM) with four different pore sizes (400, 650, 850, 1000 μm), finding that as pore size increased it favored permeability, allowing fluid entry and flow velocity. However, in vitro, results showed that a large pore size allowed cell penetration while smaller pore sizes favored cell deposition. Finally, during in vivo tests, they found that a medium pore size (650 μm) better supports bone regeneration than large pore sizes. Chen et al. [17] corroborated that pore size influences cell proliferation, osteogenesis, and bone growth. They found that smaller pore size and volume pore fraction results in better cell adhesion, proliferation, and osteogenic differentiation. Vega et al. [18] developed Ti coatings that were applied to cylindrical Ti64 endomedullary systems for evaluation in in vivo tests. They found that roughness surfaces significantly improved the cancellous bone growth around the implant, as well as good tissue–implant fixation. Deering et al. [19] fabricated porous scaffolds of 600 and 300 μm of Ti64 through SLM and performed osseointegration tests in tibiae rabbits. Through tomography, they observed incredible bone growth not only on the outside but also on the inside of the scaffolds; they found that the optimal pore size for osseointegration is 300 μm because it presented a more significant growth in 12 weeks. Another of the most widely used techniques is selective laser sintering (SLS) since it is possible to fabricate metallic scaffolds that aid bone regeneration [20]. Because of this versatility, SLS has been used to fabricate scaffolds with mechanical properties close to those of trabecular bone [21]. On the other hand, one of the main disadvantages is that the high sintering temperatures limit the inclusion of cells and biomaterials in the scaffolds [22], as well as restrict the pore size due to the characteristics of the powders, such as particle size.

Although laser techniques are outstanding, the main disadvantage is that pores are more significant than hundreds of microns, which, as pointed out above, reduces the mechanical properties but diminishes the osseointegration abilities of implants. To overcome such disadvantages, extrusion 3D printing offers an alternative to control the porosity in the parts besides larger designed pores [23,24]. This is thanks to the process involving a sintering step, which can control interparticle porosity by the thermal cycle. Among the different extrusion 3D printing technologies, stereolithography (SLA) was one of the first printing techniques for bone engineering. This is due to its precision in producing structures with nanometer and micrometer scales, in addition to having the ability to fabricate complex shapes with internal architectures and to obtain extremely high resolution (1.2 μm). However, the number of materials used in bone engineering is limited due to viscosity, stability, and refractive index restrictions.

In some cases, cytotoxic effects may be caused by irradiation with ultraviolet light [25]. Another extrusion technique is the Fused Deposition Modeling (FDM-extrusion), in which there is a sub-classification of the same depending on the type of feed that has the equipment (pellets, filaments, or bars), but they have similar procedures. This technique has more advantages over the other technologies since it is possible to control the printing parameters, such as temperatures, nozzle anchor and angle, and printing directions, which help to manufacture and control scaffolds with the desired pore size, morphology, and interconnectivity [26]. This is why the extrusion technique can produce complex 3D parts that are difficult to fabricate for other technologies (lithography and micromachining). Studies have shown that scaffolds fabricated by FDM provide favorable biochemical and mechanical properties for bone regeneration [27–30].

Nowadays, AM has emerged to develop the regenerative medicine field that aims to repair or replace damaged tissues and organs. Concerning bone implants, bone tissue engineering seeks to restore or regenerate damaged or lost bone tissue. This approach is focused on combining materials engineering with biology to generate bone implants that can better mimic bone structure and function [31,32]. As such, "active" scaffolds have been developed to serve as structures over the which the natural bone tissues will be grown and, in some cases, this structure is absorbed by the natural bone, accelerating patient recovery [33]. To improve the scaffolds, most studies are focused on controlling the porosity, which plays the major role in the new bone growth [34]. Thanks to AM, it was possible to evaluate the effect of the pore size as well as interconnectivity to the regenerative process [35,36]. In such works, it was demonstrated that pore size has a strong influence on cell adherence and interconnectivity in cell proliferation [37]. In the same way, it was shown that large pores favor vascularity and promote high permeability, which can help pass through corporal fluids with nutrients and dispose waste from the organism [38]. Further, the porosity controls the mechanical properties discussed above, thus most research is focused on developing different pore configurations. Li et al. [15] evaluated the effect of pore shape on the cell viability of titanium scaffolds, finding that a honeycomb shape induced the highest cell regeneration and the highest mechanical strength among the different shapes tested.

According to the background above, this work focuses on fabricating tailored porous materials with hexagonal channels fabricated by extrusion 3D printing of Ti64 pellets. The effect of the size of designed channels on the densification and microstructure is analyzed by SEM and X-ray microtomography. Young's moduli are obtained through compression tests to determine the most optimal for biomedical implants. Finally, permeability is evaluated by numerical simulations using authentic 3D images.

2. Materials and Methods

2.1. Raw Material

Figure 1 shows the raw material used to fabricate the porous materials, and Figure 1a–c, illustrates the Ti64 feedstock form, which is poured into the extruder of the

AIM3D printing machine (Rostock, Germany). Pellets are approximately 1 mm in diameter and 1 mm in height, furnished by PolyMIM (Bad Sobernheim, Germany). This feedstock consists of granules containing 93.5 wt.% of Ti64 powders and two polymers, namely polyethylene glycol (PEG) and some wax (not precisely specified by the provider). PEG offers flexibility during printing due to its thermoplastic nature. Figure 1c shows how the Ti64 particles are randomly embedded in the binder. From those pellets, the extrusion is performed with a nozzle of 400 μm in diameter, and the filament obtained is shown in Figure 1d. At low magnification, Ti64 particles of different sizes are observed; however, at higher magnification, the presence of the binder around the Ti64 particles is observed as shown in Figure 1e. It can be seen that the distribution of the binder is modified from its original arrangement inside the pellets.

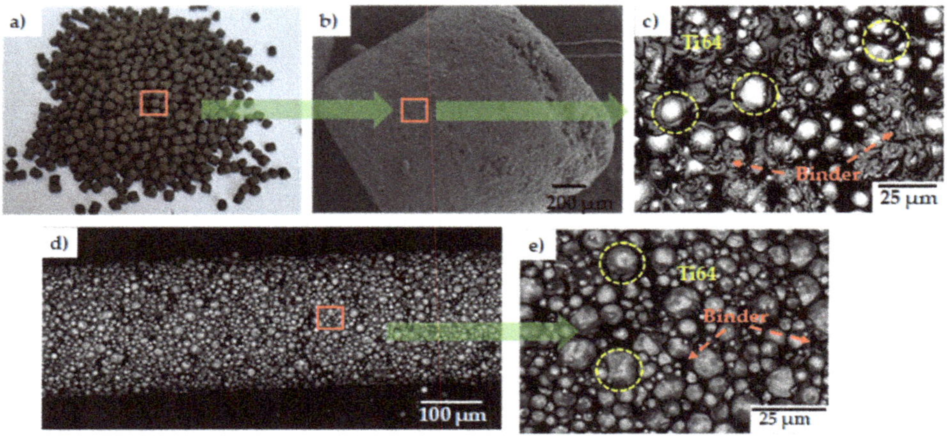

Figure 1. Ti64 raw material: (**a**) pellets, (**b**) pellet seen from SEM, (**c**) inside of pellet, (**d**) SEM micrograph of the filament obtained from print extrusion, and (**e**) larger magnification inside of filament.

2.2. Printing Parameters

Cylindrical samples of 8 mm in diameter and 10 mm in height were created in a CAD model using the SolidWorks software (2022 version). Then, the model was imported to the Simply 3D program (Version 4.0), which connects to the 3D printer. A hardened steel nozzle of 400 μm in diameter was placed at the extruder end for the filament extrusion. The impression parameters used had a layer thickness of 0.05 mm, a nozzle speed of 20 mm/s, an extrusion flow rate of 120% of the standard value, an extrusion temperature of 196 °C, and a bed temperature of 60 °C. The impression parameters were selected from previous works devoted to optimizing the full density of parts by Singh et al. [39]. They used MIM pellets of copper furnished by the same company and the same 3D printer. Additional details on the printing parameters can be found elsewhere [39–41]. To create porous samples, the Simply 3D software offers the possibility to develop trajectories that can make geometrical figures, like squares, triangles, and honeycomb shapes. Therefore, cylindrical samples were printed using the honeycomb model. A honeycomb shape is made by hexagons of the same size homogeneously distributed in the cylinder. In order to evaluate the size of the channels, honeycomb shapes were designed to be of three different sizes; from here, those samples will be called small, medium, and large. All samples were programmed to fill 80% of the total volume of cylinders, which means that channels are homogeneously distributed in the sample. Figure 2 shows the trajectories and samples designed in Simply 3D and the printed 3D cylinders. As can be seen, some defects are obtained because of the filament trajectory. However, the size of channels was different in each sample, as intended in this work.

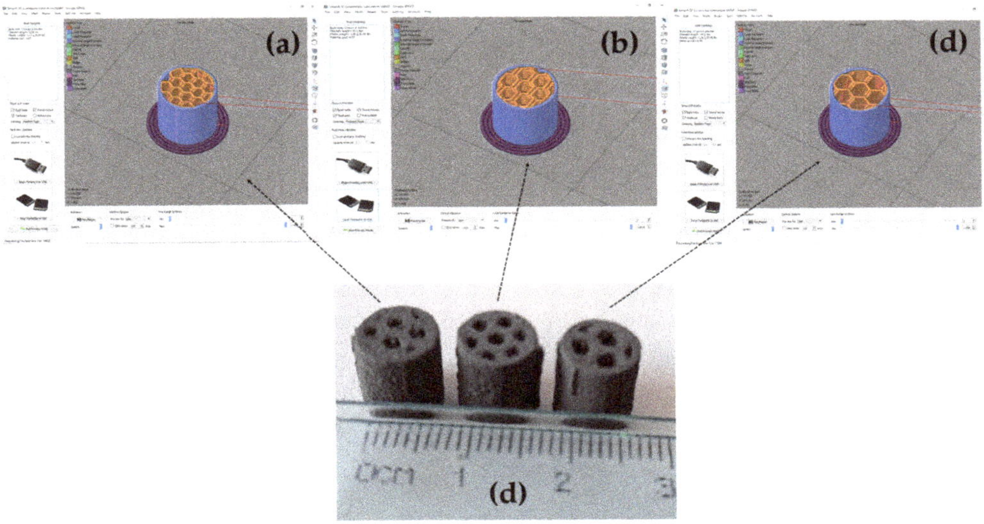

Figure 2. Images of the designed samples using the Simply 3D software: (**a**) small canals, (**b**) mediums canals, (**c**) large canals and (**d**) the resulting printed samples for each canal size.

2.3. Debinding and Sintering

After printing green parts, they must be subjected to a binder elimination process, which consists of two stages, one in solvent and the other thermal, as indicated in the literature [42]. For solvent elimination, the parts were subjected to distilled water at a temperature of 60 °C for 12 h, as indicated by Singh et al. [40], so that this would form interconnected paths throughout the sample (backbone) to help in the release of the binder (in the form of gas) during thermal elimination [43]. A Linseis L75 vertical dilatometer (Selb, Germany) with a controlled argon atmosphere was used for thermal removal. A heating ramp of 10 °C/min up to 500 °C with a dwell time of 10 min was carried out, which is supported in the study of Gonzalez et al. [44] for extrusion printed parts of similar feedstock. Consequently, within the dilatometer, the sintering process was performed, where a second ramp was added, which consisted of a heating rate of 20 °C/min up to a temperature of 1100 °C with a dwell time of 1 h. This temperature was chosen in order to control the dedensification of powders to obtain interparticle porosity (void space in between the particle packing) inside the struts. As reported before, different pore sizes help to improve the adhesion and growth of osteoblast cells, which promotes bone ingrowth [1]. Therefore, it is needed to have interconnected porosity that is reached by staying in the intermediate stage of sintering, according to the sintering theory [45]. Similar thermal cycles were used for solid samples by Su et al. [46] for Ti64 parts manufactured by MIM, in which densities of 66.59% were obtained, as well as by Cabezas et al. [47], where Ti64 parts were sintered by powder metallurgy, obtaining densities of 60%. Finally, the samples were cooled at 25 °C/min to room temperature under an argon atmosphere. Thus, the sintering thermal cycle was similar to obtain a porosity of approximately 35% inside the struts because the pellets contain Ti64 powders with a similar particle size distribution, which ensures that densification will be similar under the same sintering parameters.

2.4. Sample Characterization

3D images of printed samples were acquired before sintering with a Zeiss Xradia 510 Versa 3D X-ray microscope (Jena, Germany) with a voxel resolution of approximately 8 μm to observe the whole sample. After sintering, 3D images were also acquired using the exact voxel resolution to observe the changes in the channels. Further, samples were increased to 1.5 mm diameter, and additional 3D images were acquired with a voxel resolution of 1.5 μm to observe the microstructure at the particle scale.

After that, the porous samples were cut in half and subjected to metallographic preparations. The surface was roughened using SiC paper in ascending order, followed by mirror polishing with alumina powders with a particle size of 1 μm. Finally, they were subjected to an ultrasonic bath for 30 min to remove any residue or impurities. The microstructure was then observed with a JEOL JSM-5910LV scanning electron microscope (Tokio, Japan).

The 3D image process was carried out using the Avizo® software (V9.0) and the Image J software (https://imagej.net/ij/, accessed on 12 January 2024). The solid part channels were segmented using a thresholding procedure based on the gray level of the voxels, which is determined by the absorption coefficient of each phase [48]. This procedure is detailed by Olmos et al. [49]. The same thresholding process was performed for each channel separately (in the samples before and after sintering) in 2D to obtain the area in each channel before and after sintering. Through filters (opening and closing), the images of each channel were obtained, eliminating separate objects that were not part of the channels. Finally, the area was calculated along the sample with a label analysis where the calculation of the area in each of the slices of the image is indicated.

The numerical simulations of permeability were performed using the Avizo software in the 3D images of samples. Two different runs were performed, one in the channels of the sample and the second in the solid part, taking into account interparticle porosity. Human blood viscosity was used at 0.045 Pa s to make a more realistic simulation. The solution was obtained by solving the Navier–Stokes equations described elsewhere [50].

2.5. Mechanical Properties

Compression tests were conducted using an Instron 1150 universal machine, which provided a strain rate of 0.5 mm/s, as indicated in ASTM-D695-02 [51]. The elastic limit and Young's modulus (E) were calculated from the stress–strain data obtained by the equipment. Three different samples were tested, and the average values were taken.

3. Results

3.1. Sintering Analysis

Dilatometry was used to obtain axial deformation during the sintering thermal cycle for each sample with different channel sizes. Figure 3 depicts the deformation as a function of temperature; at the beginning of the process, a constant behavior is observed for each sample. However, as the temperature increases, a negative axial deformation (shrinkage) is observed from 140 °C, which is associated with the elimination of the binder up to 240 °C, indicating the total elimination of the binder. Then, with the temperature increase, no dimensional changes are observed for any sample up to a temperature of 650 °C, which indicates the beginning of sintering. During the sintering stage, continuous shrinkage is undergone until the isothermal period is reached. A linear shrinkage is noticed during this stage because it is obtained at a constant temperature. Finally, a final shrinkage is observed due to thermal contraction during cooling. A higher deformation is obtained for the sample with a medium channel size, followed by the large and small channels.

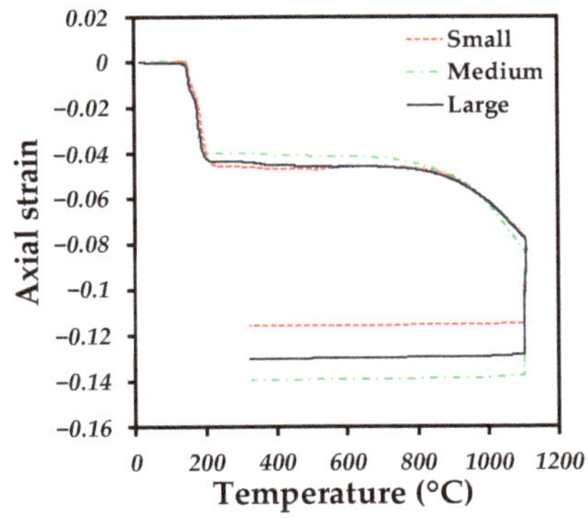

Figure 3. Axial strain as a function of temperature during the sintering cycle for samples printed with different sizes of honeycomb channels.

To analyze sintering behavior, the strain rate is plotted as a function of temperature in Figure 4. The binder elimination and the sintering steps were plotted separately in Figure 4a,b. The behavior during the thermal elimination of the binder is similar for all samples, which shows two prominent negative peaks. This confirms that the binder is eliminated in two steps, as reported before [44]. The first suggests the moment at which the binder is eliminated. This generates more extensive paths to eliminate the gas generated by the binder. The second peak indicates the total elimination of the larger binder, indicating that more binder is eliminated at the second stage, as shown in Figure 4a; after this, the strain rate is null, indicating that the dilatometer detects no dimensional changes.

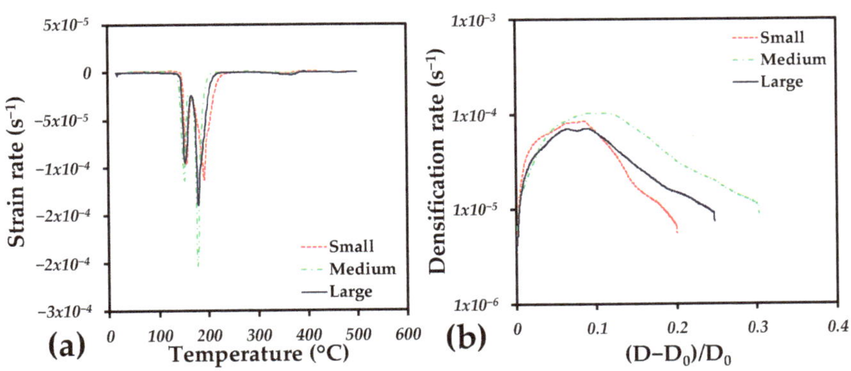

Figure 4. (**a**) The strain rate as a function of temperature during binder removal, and (**b**) the densification rate as a function of densification during the isothermal sintering period.

In order to perform a better comparison among the different samples during sintering, the densification of samples has to be estimated since they differ in initial relative density. As such, the relative density during the whole sintering cycle is obtained based on linear shrinkage as follows:

The global density is determined as:

$$\rho = \frac{m}{V}, \tag{1}$$

where ρ is the compact's density, m is mass, and is volume. Assuming that m is constant after binder elimination and during the whole densification due to sintering, the instantaneous density can be calculated as:

$$\rho_i = \frac{m}{V_i}, \tag{2}$$

where the subscript "i" means instantaneous density and volume, respectively. m is the mass measured in the sintered sample. Therefore, the instantaneous volume varies as the densification evolves during sintering, and it can be estimated from the dilatometry data. In this, the radial displacement is assumed to follow the same trend as the dilatometer measured in the axial direction. The corrective factor is the final axial-to-radial shrinkage ratio. Hence, the instantaneous volume is calculated from:

$$V_i = (\Delta l + l_o)(\Delta d + d_o)^2(\pi/4), \tag{3}$$

where Δl and Δd are determined at any instant during the thermal cycle, the relative density is defined as the compact's weight density divided by the sample's theoretical density, Ti64, with a theoretical density of 4.41 g/cm^3. Thus, the relative density (D) is calculated at any instant during the whole thermal cycle by the following equation:

$$D_i = \frac{\rho_i}{\rho_t}, \tag{4}$$

The densification of samples indicates the volume change during sintering, and it can be estimated at any instant during the whole thermal cycle as follows:

$$\frac{D_0 - D_i}{D_0}, \tag{5}$$

D_0 and D_i mean the initial and instantaneous values of the relative density. In this case, the value of D_0 is taken when the sintering temperature is reached to evaluate the isothermal stage, where most of the densification is achieved. It was found that densification is similar for the three different samples at the beginning of sintering plateau and up to 10% of the densification, as shown in Figure 4b. After that, the densification rate is faster for the sample with channels of the medium size, and slowest for the sample with the smallest channels size. In all cases, the maximum value of the densification rate is reached at 10% of the densification.

The values of relative density after printing (D_g), after the first step of binder elimination (D_b), and after sintering (D_s) are listed in Table 1. The value of D_g decreases as the channel size increases, suggesting that channel arrangement during printing allows a better distribution of the channels when they are small. This can be understood since the largest channels only have one defined channel at the center; see Figure 2. It can also be noticed that the pore volume, which is 1-D, is larger than that programmed in the printing setup. This is logical since interparticle porosity is not considered for the program. The volume fraction occupied by the channels in the sample, estimated from 3D images, confirms that samples with smaller channels are printed with high accuracy. The value of channel volume for the medium size is 21.09%, which is the closest to the intended value; as mentioned above, the printing parameters were set to fill the 80%. The sintered relative density shows the same tendency, even though the sample with channels of medium size had a larger densification during sintering.

Table 1. Relative densities of samples were measured at the different stages and after sintering.

Channel Size	(D_g) Relative Density $_{green}$ (%)	(D_b) Relative Density $_{brown}$ (%)	(D_s) Relative Density $_{sintered}$ (%)	$(D_s - D_b)/D_b$ (%)	Channel Volume Fraction (%)
Small	53.66	51.41	66.50	29.34	15.47
Medium	51.48	48.54	64.74	33.35	21.09
Large	49.81	45.91	56.44	22.94	24.77

3.2. Characterization Analysis

Figure 5 shows the microstructure of the samples after metallographic preparation. Figure 5a shows the morphology of the channels where some deformation can be seen due to the roughing process. It can be noticed from Figure 5b that there is formation of the interparticle necks, while the spherical shape of particles remains, suggesting that sintering is in the intermediate stage. Figure 5b,c shows the formation of necks between particles, generating an interparticle porosity controlled by the sintering temperature. Finally, the microstructure corresponding to the Ti64 alloy is observed (Figure 5d), showing the β phase in the form of lamellae and its more significant proportion compared to the α phase.

Figure 5. SEM images of the microstructure of sintered samples: (**a**) morphology of the channel created by printing, (**b**,**c**) necks formed between Ti64 particles, and (**d**) microstructure α and β revealed for the alloy.

Prior to the sintering process, the samples were analyzed by computed tomography; Figure 6a–c shows 3D images, where each channel size is observed, as well as its distribution throughout the sample and the morphology it has. On the other hand, Figure 6d,f shows only the channels corresponding to each size. It can be seen how the continuity of the channels is equal in the "y" direction, as well as the amount and morphology influenced by each size.

After sintering, 3D images of the same samples were obtained again, showing a slight reduction in the channel size, as can be observed qualitatively in Figure 7a–c. On the contrary, Figure 7d–e, again shows the channels corresponding to each size in which no deformation of the initial shape occurs during densification, which suggests that the channel size is homogeneous in all samples.

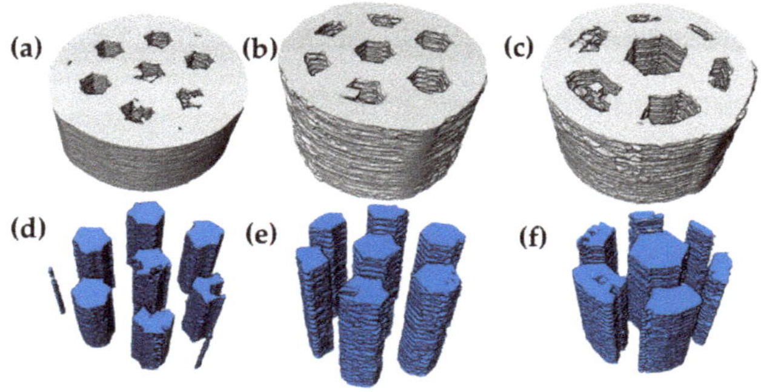

Figure 6. 3D images of printed samples before sintering: (**a**,**d**) small channels, (**b**,**e**) medium channels, and (**c**,**f**) large channels.

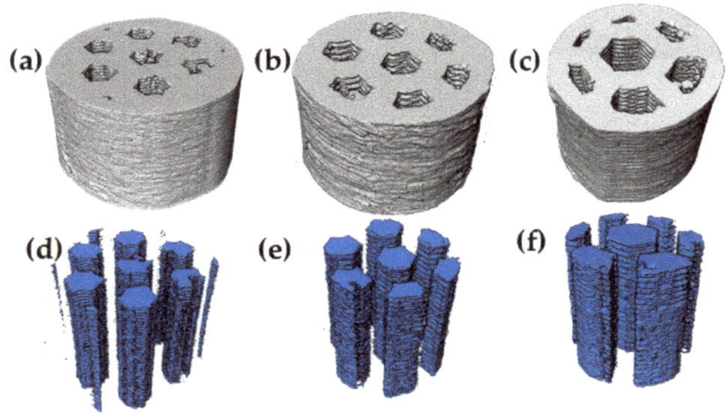

Figure 7. 3D images of printed samples after sintering: (**a**,**d**) small channels, (**b**,**e**) medium channels, and (**c**,**f**) large channels.

To quantitatively evaluate the deformation of the channels during sintering, the 2D areas before and after sintering were calculated all along the height of the sample since the same channel can be directly evaluated from 3D images.

Figure 8a–d shows the 2D images before sintering for the smaller size channels, Figure 8i–k for the medium-sized channels, and Figure 8o for the larger size channels with the measured channel indicated. The exact process was performed in the samples to compare the changes after sintering: Figure 8e–h for smaller sizes, Figure 8l,m for medium sizes, and Figure 8p for larger sizes. The area was obtained only in the channels that presented better morphology to obtain better results concerning the shape.

Table 2 lists the measured values of the areas before and after sintering; a reduction in the area value is found, which confirms that channels contributed to the densification of the sample. To estimate the change in the area, a similar value of the densification is calculated using the initial and final values of the areas A0 and As, respectively. This also confirms that the sample with medium channels underwent the most significant change in the area of the channel, confirming the tendency discussed above in the linear deformation and the densification. This suggests that channel deformation drives the densification of samples.

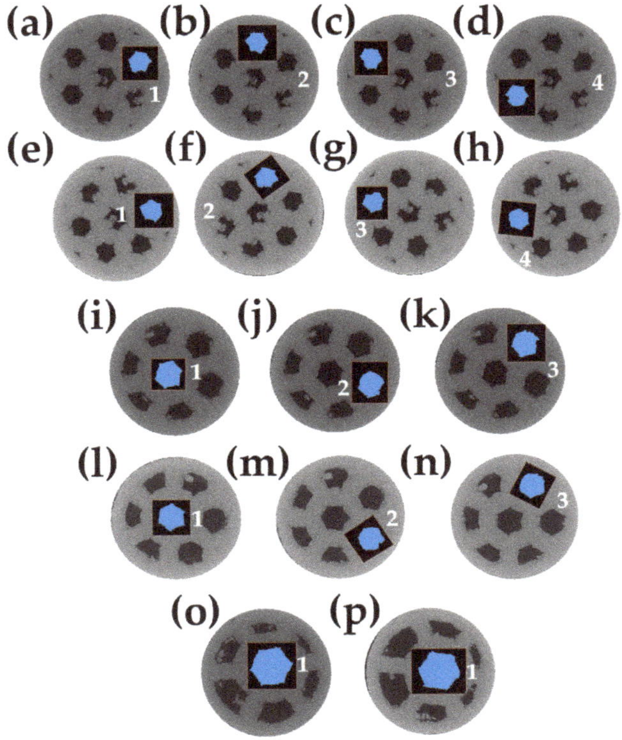

Figure 8. 2D virtual slices before and after sintering representing the channel sectioned to obtain the area: (**a–d**) small channels before sintering, (**e–h**) small channels after sintering, (**i–k**) medium channels before sintering, (**l–n**) medium channels after sintering, (**o**) large channels before sintering, and (**p**) large channels after sintering.

Table 2. Areas before and after sintering and area ratio for each of the channels contained inside the samples with different-sized channels.

Channels	Area Before Sintering (E^6)	Area After Sintering (E^6)	$(A_s–A_0)/A_0$
	Small		
Channel 1	1.22 ± 2.96	1.06 ± 4.22	−0.13436562
Channel 2	1.22 ± 2.96	1.06 ± 7.37	−0.128123664
Channel 3	1.22 ± 2.96	1.06 ± 7.37	−0.128123664
Channel 4	1.22 ± 2.96	1.06 ± 7.37	−0.128123664
	Medium		
Channel 1	1.93 ± 8.97	1.61 ± 9.62	−0.16699903
Channel 2	1.93 ± 8.96	1.61 ± 9.62	−0.16605416
Channel 3	1.93 ± 8.96	1.61 ± 9.62	−0.16605416
	Large		
Channel 1	4.50 ± 1.10	1.10 ± 1.38	−0.13248393

Figure 9 plots the area values as a function of the height of the sample before and after the sintering of one channel by the sample. It can be noticed that before sintering, small channels show more or less the same value along the sample. However, medium and large channels show more irregular area values along the sample. On the other hand, after sintering, a reduction in the area values is observed for all samples. However, the sample with small channels shows a more significant area reduction in the middle of the sample, suggesting heterogeneous deformation. This behavior is different for the sample with medium and large channels, showing a more regular value of the area values along the sample instead.

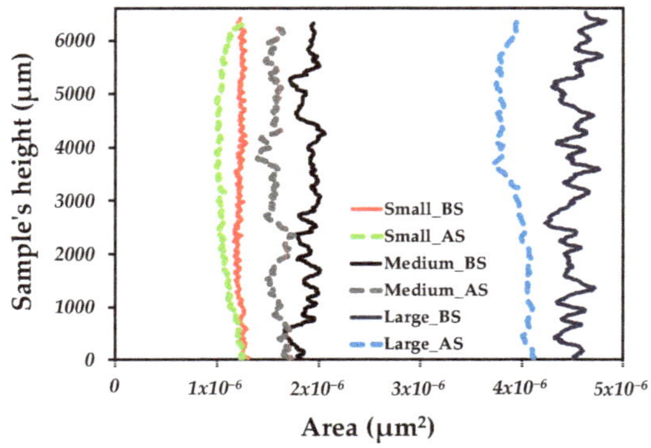

Figure 9. Areas obtained before and after sintering as a function of sample height.

3.3. Permeability Analysis

3D images were used to evaluate the permeability as a function of the channel size to determine which was the most optimal compared to human bones. As such, numerical flow simulations throughout the channels were performed using the Avizo® software. Figure 10 shows the simulated streamlined lines throughout the channels on the samples for the different samples. The flow is observed to pass only through the channels created during printing. Table 3 lists the permeability values for each sample, obtaining larger values as the channel size increases.

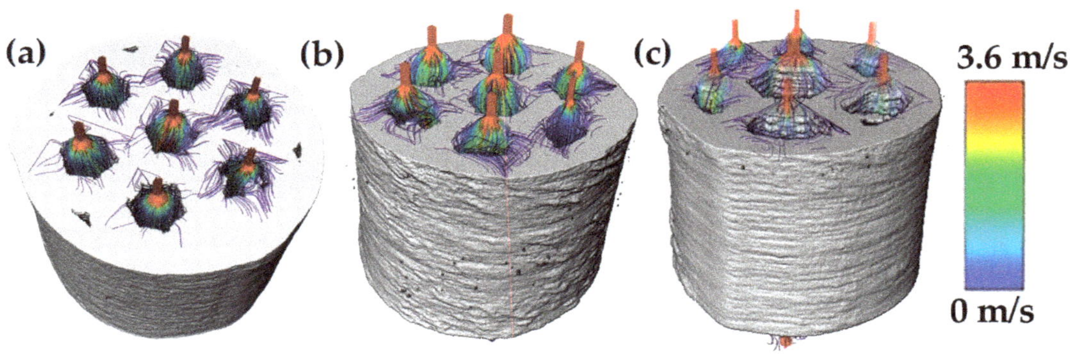

Figure 10. Simulated flow streamlines throughout the channels of different samples: (**a**) small channels, (**b**) medium channels, and (**c**) large channels.

Table 3. Permeability values and mechanical properties for samples with channels of different sizes.

Channel Size	Permeability (m^2)	E (GPa)	σ_y (MPa)
Small	0.69×10^{-8}	39.61	470.15
Medium	0.86×10^{-8}	41.74	448.60
Large	1.24×10^{-8}	37.67	394.14

Figure 11a–c shows a 2D image at high resolution after sintering acquired by computed tomography. It is observed in three principal planes to illustrate the defects that can be obtained during printing. It is found that the honeycomb shape is warped, which indicates that smaller shapes are too complex to be followed by the printer, as shown in Figure 11a. The printing traits as horizontal lines can also be noticed in Figure 11b. This is because the sintering temperature used did not allow considerable densification that typically reduces this kind of defects in fully dense parts [44]. On the other hand, a rendering was performed, as shown in Figure 11d, where the solid part of the sample and the porosity are observed; these elements are separated individually to obtain a better visualization. Figure 11e shows only the sintered Ti64 particles, showing the channel formed by the impression and interparticle porosity. Further, Figure 11f represents only the interparticle porosity, which is entirely interconnected. Figure 11e illustrates the channel created by printing, which highlights the roughness of the surface that is a consequence of the layer-by-layer printing process.

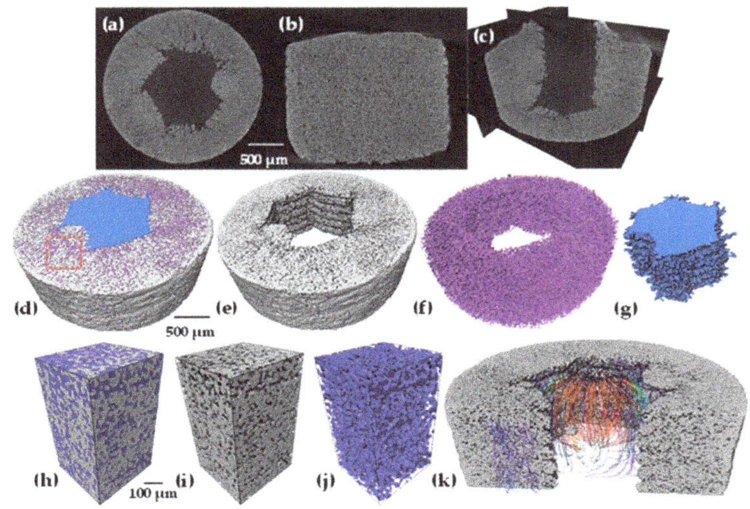

Figure 11. Virtual 2D slices showing the interparticle porosity (**a**–**c**), 3D rendering of the sample with small channels (**d**), particles (**e**), interparticle porosity (**f**), (**g**) the channel. 3D rendering inside and a virtual volume cropped (red square in (**d**)) showing the porosity (**h**–**j**) and simulated flow streamlines throughout the interparticle porosity and channel at high resolution (**k**).

A virtual volume was cropped from the 3D image to illustrate the interconnectivity of the interparticle pores. This porosity is beneficial for osseointegration and helps to achieve better anchorage between the tissue–implant interface [52]. Figure 11h–j renders this 3D volume in which the porous part is separated from the solid. Finally, permeability tests were performed in the pre-fabricated channel where the flow lines have continuity; in the same way, simulations were performed in the solid part of the material and in this way to verify the interconnectivity mentioned above.

3.4. Mechanical Properties Analysis

Figure 12 shows the stress–strain curves for all kinds of samples. In such curves, similar behavior is observed for the different samples during the elastic and plastic zones. However, they show a high resistance to failure; in the case of the large and medium-sized channel samples, they show the same behavior during the hardening zone until the test fails. On the other hand, the good ductility of samples can be observed in the images of fractured samples since they can be deformed entirely.

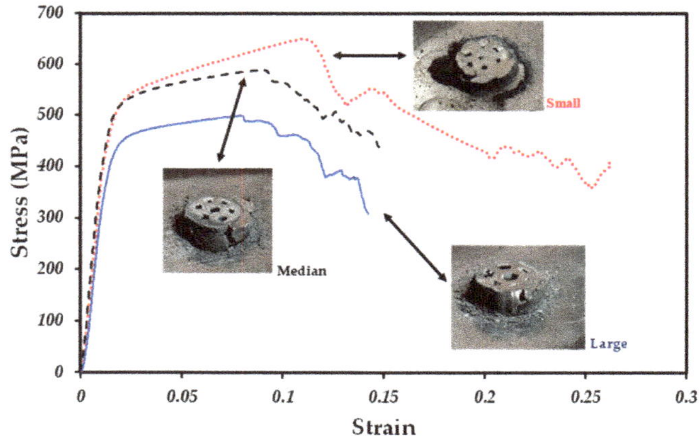

Figure 12. Stress–strain curves for samples with different channel sizes.

The yield stress (σ_y) and the Young's modulus (E) were calculated from the elastic section of each curve, and are listed in Table 3. The value of E is lower for larger channels, being 37.67 GPa. The yield values (394.14 MPa) follow the medium channel sizes, and finally, a higher value is obtained for the smaller channels, obtaining an E of 39.61 GPa and a σ_y of 470.15 MPa.

4. Discussion

Samples with two different kinds of porosity were fabricated by extrusion 3D printing, followed by debinding and the prior sintering process. Samples with channels of three different sizes were designed using a honeycomb shape. The accuracy of the samples was analyzed from 3D images, finding that the geometry is not easy to reproduce as the size is reduced. This is because the angles to be followed by the printing paths are more complicated than the traditional squares studied before [27,53,54]; however, the samples were reproducible with the same accuracy. The pore volume designed for the channels was 20%, which was not reached in all samples. The smallest channels give a lower value, but the largest ones overestimate such values. This is mainly due to the distribution of the honeycomb channels on the circular surface, which induced the difference found. According to this, the sample with channels of a medium size shows the best arrangement, reaching a 21%volume. In the vertical axis of samples, the channels were perfectly straight, confirming that each layer was accurately located one over another.

The debinding schedule followed did not generate any defect or distortion of the shape of the samples. Then, sintering was carried out at a low temperature (1100 °C) with the aim to maintain interparticle porosity, which is fully interconnected according to the sintering theory and reported in the sintering of Ti6Al4V powders with a similar particle size distribution [45,47]. It is found that the channel size has an effect on the axial shrinkage of samples. The larger axial deformation measured by the dilatometry test was found for the medium-sized channels, approximately 14%. This behavior was confirmed by analyzing the change on the surface of channels by 3D images, in which the sample with medium channels also shows the largest area reduction of 16%; see Table 2. This suggests that densification during sintering is driven by the reduction in the channel volume, which is due to the sintering stresses generated during atomic diffusion and neck formation. In a general way, sintering at the particle scale is not affected by the presence of the channels as demonstrated in SEM images, in which the interparticle necks were well developed, as shown in Figure 5. The evolution of sintering was also in the intermediate stage, with interparticle porosity fully interconnected as shown in Figure 11, which confirms the fabrication of samples with two kinds of porosities that are fully interconnected among them. The graded porosity is beneficial for improving bone ingrowth, since small pores allow cell adhesion and cell proliferation, while large pores help in the vascularization to provide the nutrients required for bone ingrowth, as was demonstrated for graded porous materials of P-PEEK material [55]. This microstructure can better mimic the function of real bones, which could help to improve the osseointegration of such materials, as also reported for Ti6Al4V scaffolds fabricated by AM and implanted in rabbits [56].

Permeability was evaluated by numerical simulations from two points of view—at the sample and particle scales. This means that first numerical simulations were run on the whole samples, in which the channels were the only paths for the liquid to pass through the sample, as illustrated in Figure 10. In the second approach, numerical simulation was run on the interparticle porosity that connects with one channel, as shown in Figure 11. The values of permeability increased as the channel size did, which follows the geometrical assumptions made by Kozeny-Carman [57] and Katz and Thompson [58], where they found that as the volume fraction of pores and the pore size increase, the permeability does so too. On the other hand, the estimated values were in the range of 6.912.4 $\times$ 10^{-9} m^2, similar to those measured for trabecular bones by Grim and Williams 0.4–11 $\times$ 10^{-9} m^2 [59]. Those values are also in the range reported for trabecular-like porous structure materials (0.6 $\times$ 10^{-9}–21 $\times$ 10^{-9} m^2) that showed good osseointegration of such materials [60]. Thereby, permeability is directly linked to the porous characteristics as well as to the interconnection in scaffolds, which is also linked to the transport of nutrient and oxygen content. This parameter is very important since it has been demonstrated that colonization by bone cells on the scaffold surface needs different permeability values; further, the largest values improve vascularization, and lower values of permeability can enhance the cartilaginous ECM production of chondrocytes [61]. Therefore, pore connectivity in the whole sample would be beneficial to improve bone ingrowth and accelerate patient recovery.

The Young's modulus (E) evaluated by compression tests shows similar values between 37 and 41 GPa for channels with different sizes; see Table 3. Those values are a bit higher to the ones reported for trabecular bones (1 to 22.3 GPa) that depend on the physiological and pathological conditions of bones [62]. They are also higher when compared to that of materials with higher pore volume fractions fabricated by similar processing techniques [29,63]. On the other hand, the compressive strength determined by the yield stress (σ_y) ranges between 394 and 470 MPa depending on the channel size of samples, as listed in Table 3. Those values are also higher in comparison to the ones reported for the trabecular bones (5–200 MPa) by Wang et al. [64]. However, the values are in a low range compared to that of the fully dense materials actually used as bone implants and as noted, those values can be reduced by increasing the pore volume fraction in order to match the trabecular bone mechanical properties.

Table 4 lists the mechanical properties reported in different works of scaffold fabricated by AM techniques with different topologies or shapes of pores. It can be seen that most works were devoted to mimic the spongy bones because the relative density of such scaffolds is lower than 40%. Thus, the mechanical properties are low in most cases and in the spongy bones range.

Table 4. Mechanical properties as a function of the relative density reported for scaffold fabricated by different AM techniques.

Topology	Relative Density (%)	E (MPa)	σ_y (MPa)	Ref.
BCC	16.5–36.3	500.0–1500.0	25.9–68.6	[65]
BCCZ	5.5–40.3	750.6–2338.6	3.58–24.7	[66]
SC	9.9–13.1	1011.6–1149.1	20.2–29.3	[67]
SC	16.6–30.4	2362.9–3793.7	75.7–158.2	[68]
SC	21.9–39.2	5270.1–7565.2	94.6–188.1	[69]
Hybrid-Strut SC	15.7–30.2	1893.3–4438.5	61.8–181.6	[68]
NR-SC	11.6–16.4	1715.8–2257.5	43.2–71.6	[67]
FCC	8.1–13.7	724.2–1271.2	15.7–36.7	[70]
FCC	15.8	1930	50.2	[71]
FCCZ	9.0–15.6	1020.1–2019.1	19.7–53.0	[70]
Octahedron	16.7	1890	49.6	[71]
Octet Truss	28.2	3380	109.5	[71]
Multi-topology SC	21.8–39.7	6376.4–10,517.0	134.3–259.9	[69]

To perform a comparison with the materials developed in this work, the Young's modulus and the yield strength were plotted as a function of the pore volume fraction of the different samples reported in the literature and the ones fabricated in this work; see Figure 13. The behavior observed for the E and σ_y shows a reduction as the pore volume increases, which is logical. In order to get an idea of the behavior, the model proposed by Gibson and Ashby [72] to predict the mechanical properties is also plotted in Figure 13. This model suggests that the E value is reduced in a power law as follows:

$$E = E_0(1-p)^2 \tag{6}$$

where the E and E_0 are the Young's modulus of the porous material and the material without pores—for this case, 110 GPa was used, which corresponds to the value of the Ti6Al4V solid alloy. Further, p is the pore volume fraction of samples. It was found that the Gibson and Ashby model shows good fit with the experimental values measured for the E, those of this work and those reported in the literature; Figure 13a. Similarly, the model proposed for the σ_y for porous materials is similar, but with an exponential a little lower than 2:

$$\sigma = E\sigma_0(1-p)^{1.5} \tag{7}$$

where the σ and σ_0 are the yield strength of the porous material and the material without pores—for this case, 840 MPa was used, which corresponds to the value of the Ti6Al4V solid alloy. As shown in Figure 13b, the model can also predict the values reported very well. For both properties, the highest values reported for human bones are reported, which are lower than those obtain in this work. Nevertheless, it is also noticed that the porosity drives the mechanical properties, so, in order to reach similar values to human bones, additional porosity should be added to the materials fabricated by extrusion 3D printing. This will also provide an improvement in the permeability, which will be beneficial for the osseointegration functions and cell adhesion will always be maintained for the interparticle porosity inside the struts. This kind of porosity can be only obtained for AM techniques that involve extrusion of pellets or filaments, rather than laser techniques.

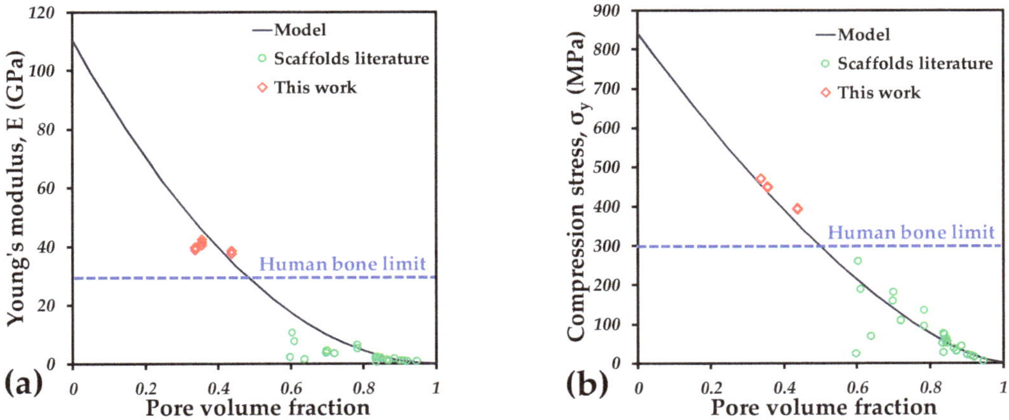

Figure 13. Mechanical properties of scaffolds fabricated by AM as a function of the porosity: (**a**) Young's modulus and (**b**) yield stress.

5. Conclusions

Cylindrical samples with honeycomb-shaped channels of different sizes were designed and fabricated by extrusion 3D printing using Ti6Al4V MIM feedstock. The shape was not precisely obtained because the angles to form the honeycomb shape are complex for the 3D printer. In spite of that, samples were accurately printed in the vertical axis, keeping the shape of channels very similar along the sample. No defects or fissures were found after the debinding and sintering processes. The main objective of this work on developing materials with two kinds of porosities was achieved, with fully interconnected channels and interparticle pores. Permeability values are in the range of trabecular bones, which is promising for the osseointegration of such materials. The mechanical properties are a little higher in comparison to that of trabecular bones; nevertheless, these properties can be tuned by increasing the pore volume fraction during the printing stage. The main conclusion is that materials with two scales of porosity can be developed by the 3D printing extrusion process.

Evaluation of cell attachment and bone ingrowth with this kind of sample is in progress to determine what size will be the most effective. This will lead to the fabrication of bone implants like hip, knees or vertebral implants, with the best size and configuration of channels and tailored porosity according to requirements.

Author Contributions: L.O., writing—original draft preparation, conceptualization, and formal analysis; A.S.G.-P., writing—original draft preparation, methodology, and data curation; H.J.V.-H., conceptualization, funding acquisition, and resources; D.B., supervision and project management; M.S.L.-C., writing—editing and methodology; R.S.-C., investigation and data curation. All authors have read and agreed to the published version of the manuscript.

Funding: This research was supported by the National Council of Humanities, Science and Technology CONAHCYT via the PhD scholarship of A.S. Gonzaléz-Pedraza, grant number [CVU 1087636].

Institutional Review Board Statement: Not applicable.

Informed Consent Statement: Not applicable.

Data Availability Statement: The data presented in this study are available on request from the corresponding author due to the raw/processed data required to reproduce these findings cannot be shared at this time as the data also form part of an ongoing study.

Acknowledgments: The authors would like to thank to the CIC of the UMSNH for the financial support and the facilities provided for the development of this study. A special acknowledgment to Dante Arteaga for 3D image acquisition and processing performed at the Laboratory "LUMIR" Geosciences of the UNAM, Juriquilla, Queretaro, Mexico.

Conflicts of Interest: The authors declare no conflict of interest.

References

1. Lutzweiler, G.; Ndreu Halili, A.; Engin Vrana, N. The overview of porous, bioactive scaffolds as instructive biomaterials for tissue regeneration and their clinical translation. *Pharmaceutics* **2020**, *12*, 602. [CrossRef]
2. Yu, G.; Li, Z.; Li, S.; Zhang, Q.; Hua, Y.; Liu, Y.; Zhao, X. The select of internal architecture for porous Ti alloy scaffold: A compromise between mechanical properties and permeability. *Mater. Des.* **2020**, *192*, 108754. [CrossRef]
3. Fermenia, J.L.; Borrás, V.A. *Desarrollo de Aleaciones de Titanio: Implicaciones en el Sector Biomédico Europeo*; Universitat Politecnica de Valencia: Valencia, Spain, 2023. [CrossRef]
4. Antunes, L.H.M.; De Lima, C.R.P. Cobalt-Chromium Alloys—Properties and Applications. In *Reference Module in Materials Science and Materials Engineering*; Elsevier: Amsterdam, The Netherlands, 2018; pp. 1–4. [CrossRef]
5. Ali, S.; Irfan, M.; Niazi, U.M.; Rani, A.M.A.; Rashedi, A.; Rahman, S.; Khan, M.K.A.; Alsaiari, M.A.; Legutko, S.; Petrů, J.; et al. Microstructure and Mechanical Properties of Modified 316L Stainless Steel Alloy for Biomedical Applications Using Powder Metallurgy. *Materials* **2022**, *15*, 2822. [CrossRef] [PubMed]
6. Adamovic, D.; Ristic, B.; Zivic, F. Review of existing biomaterials—Method of material selection for specific applications in orthopedics. In *Biomaterials in Clinical Practice: Advances in Clinical Research and Medical Devices*; Springer Nature: Berlin/Heidelberg, Germany, 2018; pp. 47–99. [CrossRef]
7. Yaqoob, K.; Amjad, I.; Awan, M.A.M.; Liaqat, U.; Zahoor, M.; Kashif, M. Novel Method for the Production of Titanium Foams to Reduce Stress Shielding in Implants. *ACS Omega* **2023**, *8*, 1876–1884. [CrossRef]
8. Alkentar, R.; Kladovasilakis, N.; Tzetzis, D.; Mankovits, T. Effects of pore size parameters of titanium additively manufactured lattice structures on the osseointegration process in orthopedic applications: A comprehensive review. *Crystals* **2023**, *13*, 113. [CrossRef]
9. Hassan, A.; Alnaser, I.A. A Review of Different Manufacturing Methods of Metallic Foams. *ACS Omega* **2024**, *9*, 6280–6295. [CrossRef]
10. Olmos, L.; González-Pedraza, A.S.; Vergara-Hernández, H.J.; Chávez, J.; Jiménez, O.; Mihalcea, E.; Arteaga, D.; Ruiz-Mondragón, J.J. Ti64/20Ag Porous Composites Fabricated by Powder Metallurgy for Biomedical Applications. *Materials* **2022**, *15*, 5956. [CrossRef] [PubMed]
11. Garnica, P.; Macias, R.; Chavez, J.; Bouvard, D.; Jimenez, O.; Olmos, L.; Arteaga, D. Fabrication and characterization of highly porous Ti6Al4V/xTa composites for orthopedic applications. *J. Mater. Sci.* **2020**, *55*, 16419–16431. [CrossRef]
12. Aguilar, C.; Martin, F.S.; Cámara, B.; Claverías, F.; Undabarrena, A.; Sancy, M.; Salinas, V.; Muñoz, L. Improving mechanical properties and antibacterial response of α/β ternary Ti–Ta alloy foams for biomedical uses. *J. Mater. Res. Technol.* **2023**, *24*, 8735–8753. [CrossRef]
13. Karakurt, E.M.; Cetin, Y.; Incesu, A.; Demirtas, H.; Kaya, M.; Yildizhan, Y.; Tosun, M.; Huang, Y. Microstructural, Biomechanical, and In Vitro Studies of Ti-Nb-Zr Alloys Fabricated by Powder Metallurgy. *Materials* **2023**, *16*, 4240. [CrossRef] [PubMed]
14. Gao, L.-Y.; Yang, H.-K.; Chen, X.; Tang, W.-D.; Huang, X.-M.; Liu, Z.-Q. The development of porous metallic materials: A short review of fabrication, characteristics, and applications. *Phys. Scr.* **2023**, *98*, 122001. [CrossRef]
15. Li, B.; Hesar, B.D.; Zhao, Y.; Ding, L. Design and additive manufacturing of porous titanium scaffolds for optimum cell viability in bone tissue engineering. *Proc. Inst. Mech. Eng. Part B J. Eng. Manuf.* **2023**, *237*, 2015–2024. [CrossRef]
16. Ouyang, P.; Dong, H.; He, X.; Cai, X.; Wang, Y.; Li, J.; Li, H.; Jin, Z. Hydromechanical mechanism behind the effect of pore size of porous titanium scaffolds on osteoblast response and bone ingrowth. *Mater. Des.* **2019**, *183*, 108151. [CrossRef]
17. Chen, Z.; Yan, X.; Yin, S.; Liu, L.; Liu, X.; Zhao, G.; Ma, W.; Qi, W.; Ren, Z.; Liao, H.; et al. Influence of the pore size and porosity of selective laser melted Ti6Al4V ELI porous scaffold on cell proliferation, osteogenesis and bone ingrowth. *Mater. Sci. Eng.* **2020**, *106*, 110289. [CrossRef] [PubMed]
18. Vega, C.A.; Moruno, F.; Veneri, D. Rugosidad óptima de un implante para generar la osteointegración y la fijación biológica. *Rev. Asoc. Argent. Ortop. Traumatol.* **2024**, *89*, 507–518. [CrossRef]
19. Deering, J.; Mahmoud, D.; Rier, E.; Lin, Y.; do Nascimento Pereira, A.C.; Titotto, S.; Fang, Q.; Wohl, G.R.; Deng, F.; Grandfield, K.; et al. Osseointegration of functionally graded Ti6Al4V porous implants: Histology of the pore network. *Biomater. Adv.* **2023**, *155*, 213697. [CrossRef] [PubMed]
20. Bae, E.J.; Kim, J.H.; Kim, W.C.; Kim, H.Y. Bond and fracture strength of metal-ceramic restorations formed by selective laser sintering. *J. Adv. Prosthodont.* **2014**, *6*, 266–271. [CrossRef]

21. Liao, H.T.; Lee, M.Y.; Tsai, W.W.; Wang, H.C.; Lu, W.C. Osteogenesis of adipose-derived stem cells on polycaprolactone–β-tricalcium phosphate scaffold fabricated via selective laser sintering and surface coating with collagen type I. *J. Tissue Eng. Regen. Med.* **2016**, *10*, E337–E353. [CrossRef]
22. Mazzoli, A. Selective laser sintering in biomedical engineering. *Med. Biol. Eng. Comput.* **2013**, *51*, 245–256. [CrossRef] [PubMed]
23. Tejo-Otero, A.; Buj-Corral, I.; Fenollosa-Artés, F. 3D printing in medicine for preoperative surgical planning: A review. *Ann. Biomed. Eng.* **2020**, *48*, 536–555. [CrossRef]
24. Zhang, L.; Yang, G.; Johnson, B.N.; Jia, X. Three-dimensional (3D) printed scaffold and material selection for bone repair. *Acta Biomater.* **2019**, *84*, 16–33. [CrossRef] [PubMed]
25. Melchels, F.P.; Feijen, J.; Grijpma, D.W. A review on stereolithography and its applications in biomedical engineering. *Biomaterials* **2010**, *31*, 6121–6130. [CrossRef]
26. Wu, W.; Geng, P.; Li, G.; Zhao, D.; Zhang, H.; Zhao, J. Influence of layer thickness and raster angle on the mechanical properties of 3D-printed PEEK and a comparative mechanical study between PEEK and ABS. *Materials* **2015**, *8*, 5834–5846. [CrossRef]
27. Putra, N.E.; Leeflang, M.A.; Minneboo, M.; Taheri, P.; Fratila-Apachitei, L.E.; Mol, J.M.C.; Zhou, J.; Zadpoor, A.A. Extrusion-based 3D printed biodegradable porous iron. *Acta Biomater.* **2021**, *121*, 741–756. [CrossRef]
28. Sa, M.W.; Nguyen, B.N.B.; Moriarty, R.A.; Kamalitdinov, T.; Fisher, J.P.; Kim, J.Y. Fabrication and evaluation of 3D printed BCP scaffolds reinforced with ZrO_2 for bone tissue applications. *Biotechnol. Bioeng.* **2018**, *115*, 989–999. [CrossRef]
29. Slámečka, K.; Kashimbetova, A.; Pokluda, J.; Zikmund, T.; Kaiser, J.; Montufar, E.B.; Čelko, L. Fatigue behaviour of titanium scaffolds with hierarchical porosity produced by material extrusion additive manufacturing. *Mater. Des.* **2023**, *225*, 111453. [CrossRef]
30. Coffigniez, M.; Gremillard, L.; Boulnat, X. Sinter-Based Additive Manufacturing of Graded Porous Titanium Scaffolds by Multi-Inks 3D Extrusion. *Adv. Eng. Mater.* **2023**, *25*, 2201159. [CrossRef]
31. Kantaros, A.; Piromalis, D. Fabricating lattice structures via 3D printing: The case of porous bio-engineered scaffolds. *Appl. Mech.* **2021**, *2*, 289–302. [CrossRef]
32. Kantaros, A. 3D printing in regenerative medicine: Technologies and resources utilized. *Int. J. Mol. Sci.* **2022**, *23*, 14621. [CrossRef] [PubMed]
33. Tang, G.; Liu, Z.; Liu, Y.; Yu, J.; Wang, X.; Tan, Z.; Ye, X. Recent trends in the development of bone regenerative biomaterials. *Front. Cell Dev. Biol.* **2021**, *9*, 665813. [CrossRef] [PubMed]
34. Flores-Jiménez, M.S.; Garcia-Gonzalez, A.; Fuentes-Aguilar, R.Q. Review on porous scaffolds generation process: A tissue engineering approach. *ACS Appl. Bio Mater.* **2023**, *6*, 1–23. [CrossRef] [PubMed]
35. Doryab, A.; Schmid, O. Bioactive cell-derived ECM scaffold forms a unique cellular microenvironment for lung tissue engineering. *Biomedicines* **2022**, *10*, 1791. [CrossRef]
36. Ghorbani, F.; Zamanian, A.; Kermanian, F.; Shamoosi, A. A bioinspired 3D shape olibanum-collagen-gelatin scaffolds with tunable porous microstructure for efficient neural tissue regeneration. *Biotechnol. Prog.* **2020**, *36*, e2918. [CrossRef]
37. Matsiko, A.; Gleeson, J.P.; O'Brien, F.J. Scaffold Mean Pore Size Influences Mesenchymal Stem Cell Chondrogenic Dierentiation and Matrix Deposition. *Tissue Eng. Part A* **2014**, *21*, 486–497. [CrossRef]
38. Murphy, W.L.; Dennis, R.G.; Kileny, J.L.; Mooney, D.J. Salt fusion: An approach to improve pore interconnectivity within tissue engineering scaffolds. *Tissue Eng.* **2002**, *8*, 43–52. [CrossRef]
39. Singh, G.; Missiaen, J.M.; Bouvard, D.; Chaix, J.M. Copper extrusion 3D printing using metal injection moulding feedstock: Analysis of process parameters for green density and surface roughness optimization. *Addit. Manuf.* **2021**, *38*, 101778. [CrossRef]
40. Singh, G.; Missiaen, J.M.; Bouvard, D.; Chaix, J.M. Additive manufacturing of 17–4 PH steel using metal injection molding feedstock: Analysis of 3D extrusion printing, debinding and sintering. *Addit. Manuf.* **2021**, *47*, 102287. [CrossRef]
41. Singh, G.; Missiaen, J.M.; Bouvard, D.; Chaix, J.M. Copper additive manufacturing using MIM feedstock: Adjustment of printing, debinding, and sintering parameters for processing dense and defectless parts. *Int. J. Adv. Manuf. Technol.* **2021**, *115*, 449–462. [CrossRef]
42. Lotfizarei, Z.; Mostafapour, A.; Barari, A.; Jalili, A.; Patterson, A.E. Overview of debinding methods for parts manufactured using powder material extrusion. *Addit. Manuf.* **2023**, *61*, 103335. [CrossRef]
43. Hamidi, M.F.F.A.; Harun, W.S.W.; Khalil, N.Z.; Ghani, S.A.C.; Azir, M.Z. Study of solvent debinding parameters for metal injection moulded 316L stainless steel. *IOP Conf. Ser. Mater. Sci. Eng.* **2017**, *257*, 012035. [CrossRef]
44. González-Pedraza, A.S.; Bouvard, D.; Missiaen, J.M.; Olmos, L.; Vergara-Hernández, H.J.; Lemus-Ruiz, J.; Villalobos, J.C. Analyzing the Debinding Step of Ti64 Parts Fabricated by 3D Printing Extrusion. *Coatings* **2024**, *14*, 715. [CrossRef]
45. German, R.M. *Sintering Theory and Practice*; John Willey & Sons: Hoboken, NJ, USA, 1996.
46. Su, S.; Hong, Z.; Huang, Y.; Liu, Y.; Li, X.; Wu, J.; Wu, Y. Preparation, characterization and designing of porous Ti–6Al–4V through metal injection molding without space holder. *J. Mater. Res. Technol.* **2023**, *27*, 5651–5661. [CrossRef]
47. Cabezas-Villa, J.L.; Lemus-Ruiz, J.; Bouvard, D.; Jiménez, O.; Vergara-Hernández, H.J.; Olmos, L. Sintering study of Ti6Al4V powders with different particle sizes and their mechanical properties. *Int. J. Miner. Metall. Mater.* **2018**, *25*, 1389–1401. [CrossRef]

48. Olmos, L.; Takahashi, T.; Bouvard, D.; Martin, C.L.; Salvo, L.; Bellet, D.; Di Michiel, M. Analysing the sintering of heterogeneous powder structures by in situ microtomography. *Philos. Mag.* **2009**, *89*, 2949–2965. [CrossRef]
49. Olmos, L.; Bouvard, D.; Cabezas-Villa, J.L.; Lemus-Ruiz, J.; Jiménez, O.; Arteaga, D. Analysis of compression and permeability behavior of porous Ti6Al4V by computed microtomography. *Met. Mater. Int.* **2019**, *25*, 669–682. [CrossRef]
50. Callow, B.; Falcon-Suarez, I.; Marin-Moreno, H.; Bull, J.M.; Ahmed, S. Optimal X-ray micro-CT image based methods for porosity and permeability quantification in heterogeneous sandstones. *Geophys. J. Int.* **2020**, *223*, 1210–1229. [CrossRef]
51. *ASTM D695-02*; Standard Test Method for Compressive Properties of Rigid Plastics. ASTM International: West Conshohocken, PA, USA, 2020. [CrossRef]
52. Wang, D.; He, G.; Tian, Y.; Ren, N.; Ni, J.; Liu, W.; Zhang, X. Evaluation of channel-like porous-structured titanium in mechanical properties and osseointegration. *J. Mater. Sci. Technol.* **2020**, *44*, 160–170. [CrossRef]
53. Jakus, A.E.; Geisendorfer, N.R.; Lewis, P.L.; Shah, R.N. 3D-printing porosity: A new approach to creating elevated porosity materials and structures. *Acta Biomater.* **2018**, *72*, 94–109. [CrossRef] [PubMed]
54. Xu, C.; Qi, J.; Zhang, L.; Liu, Q.; Ren, L. Material extrusion additive manufacturing of Ti6Al4V bio-inspired bone implants with tunable Young's modulus. *Addit. Manuf.* **2023**, *78*, 103884. [CrossRef]
55. Wang, D.; Jiang, C.; Li, J.; Guo, J.; Zhang, J.; Ba, F.; Li, Y.; He, G. Effects of porosity distribution on mechanical properties and osseointegration of porous polyetheretherketone. *Biomater. Adv.* **2025**, *166*, 214043. [CrossRef] [PubMed]
56. Jacobs, Z.; Schipani, R.; Pastrama, M.; Ahmadi, S.M.; Sajadi, B. Evaluation of biocompatibility and osseointegration of multi-component TiAl6V4 titanium alloy implants. *J. Orthop. Res.* **2025**, *43*, 139–152. [CrossRef]
57. Schulz, R.; Ray, N.; Zech, S.; Rupp, A.; Knabner, P. Beyond Kozeny–Carman: Predicting the permeability in porous media. *Transp. Porous Media* **2019**, *130*, 487–512. [CrossRef]
58. Katz, A.J.; Thompson, A.H. Quantitative prediction of permeability in porous rock. *Phys. Rev. B* **1986**, *34*, 8179. [CrossRef] [PubMed]
59. Grimm, M.J.; Williams, J.L. Measurements of permeability in human calcaneal trabecular bone. *J. Biomech.* **1997**, *30*, 743–745. [CrossRef] [PubMed]
60. Chao, L.; He, Y.; Gu, J.; Xie, D.; Yang, Y.; Shen, L.; Wu, G.; Wang, L.; Tian, Z. Evaluation of compressive and permeability behaviors of trabecular-like porous structure with mixed porosity based on mechanical topology. *J. Funct. Biomater.* **2023**, *14*, 28. [CrossRef] [PubMed]
61. Kemppainen, J.M.; Hollister, S.J. Differential effects of designed scaffold permeability on chondrogenesis by chondrocytes and bone marrow stromal cells. *Biomaterials* **2010**, *31*, 279–287. [CrossRef]
62. Wu, D.; Isaksson, P.; Ferguson, S.J.; Persson, C. Young's modulus of trabecular bone at the tissue level: A review. *Acta Biomater.* **2018**, *78*, 1–12. [CrossRef]
63. Wang, Z.; Zhang, M.; Liu, Z.; Wang, Y.; Dong, W.; Zhao, S.; Sun, D. Biomimetic design strategy of complex porous structure based on 3D printing Ti-6Al-4V scaffolds for enhanced osseointegration. *Mater. Des.* **2022**, *218*, 110721. [CrossRef]
64. Wang, X.; Xu, S.; Zhou, S.; Xu, W.; Leary, M.; Choong, P.; Qian, M.; Brandt, M.; Xie, Y.M. Topological design and additive manufacturing of porous metals for bone scaffolds and orthopaedic implants: A review. *Biomaterials* **2016**, *83*, 127–141. [CrossRef]
65. Zhong, H.Z.; Song, T.; Li, C.W.; Das, R.; Gu, J.F.; Qian, M. Understanding the superior mechanical properties of hollow-strut metal lattice materials. *Scr. Mater.* **2023**, *228*, 115341. [CrossRef]
66. Noronha, J.; Leary, M.; Qian, M.; Kyriakou, E.; Brandt, M. Geometrical parameters and mechanical properties of Ti6Al4V hollow-walled lattices. *Mater. Sci. Eng. A* **2022**, *840*, 142667. [CrossRef]
67. Noronha, J.; Dash, J.; Leary, M.; Downing, D.; Kyriakou, E.; Brandt, M.; Qian, M. Node-reinforced hollow-strut metal lattice materials for higher strength. *Scr. Mater.* **2023**, *234*, 115547. [CrossRef]
68. Noronha, J.; Dash, J.; Downing, D.; Rogers, J.; Qian, M.; Brandt, M.; Leary, M. Ti–6Al–4V hybrid-strut lattice metamaterials: A design strategy for improved performance. *Mater. Sci. Eng. A* **2024**, *911*, 146918. [CrossRef]
69. Noronha, J.; Qian, M.; Leary, M.; Kyriakou, E.; Brudler, S.; Brandt, M. Manufacturability of Ti-Al-4V hollow-walled lattice struts by laser powder bed fusion. *JOM* **2021**, *73*, 4199–4208. [CrossRef]
70. Noronha, J.; Rogers, J.; Leary, M.; Kyriakou, E.; Inverarity, S.B.; Das, R.; Brandt, M.; Qian, M. Ti-6Al-4V hollow-strut lattice materials by laser powder bed fusion. *Addit. Manuf.* **2023**, *72*, 103637. [CrossRef]
71. Zhang, Z.; Zhang, L.; Song, B.; Yao, Y.; Shi, Y. Bamboo-inspired, simulation-guided design and 3D printing of light-weight and high-strength mechanical metamaterials. *Appl. Mater. Today* **2022**, *26*, 101268. [CrossRef]
72. Gibson, L.J.; Ashby, M.F. *Cellular Solids: Structure and Properties*, 2nd ed.; Cambridge University Press: Cambridge, UK, 1999; Volume 175.

materials

MDPI

Article

Investigation of Pore Size Effect on the Infiltration Process of Ti6Al4V/xAg Metal Matrix Composites

Juan Israel Villa-Tapia [1], Héctor Javier Vergara-Hernández [1], Luis Olmos [2], Dante Arteaga [3], Jorge Sergio Téllez-Martínez [1], Víctor Manuel Solorio-García [1] and Elena Mihalcea [4,*]

1 División de Estudios de Posgrado e Investigación, Tecnológico Nacional de México/I.T. Morelia, Av. Tecnológico #1500, Colonia Lomas de Santiaguito, Morelia C.P. 58120, Mexico; m14120336@morelia.tecnm.mx (J.I.V.-T.); hector.vh@morelia.tecnm.mx (H.J.V.-H.); jstellezm@gmail.com (J.S.T.-M.); v_ms_g@hotmail.com (V.M.S.-G.)

2 Instituto de Investigaciones en Ciencias de la Tierra (INICIT), Universidad Michoacana de San Nicolás de Hidalgo, Fco. J.Mujica S/N, Morelia C.P. 58060, Mexico; luis.olmos@umich.mx

3 Centro de Geociencias, Universidad Nacional Autónoma de México, Blvd. Juriquilla No. 3001, Querétaro C.P. 76230, Mexico; darteaga@geociencias.unam.mx

4 Unidad Académica de Ingeniería I, Universidad Autónoma de Zacatecas, Av. López Valverde 801, Zacatecas C.P. 98060, Mexico

* Correspondence: emihalcea@itmorelia.edu.mx

Abstract: This work investigates the fabrication of Ti6Al4V composites manufactured by powder metallurgy through pressureless infiltration. Porous Ti6Al4V alloy compacts with different particle sizes were fabricated by sintering and then, liquid Ag was infiltrated to obtain composites. Computed microtomography was used to analyze the samples before and after infiltration. Numerical flow simulations and dilatometry tests evaluated the kinetics of Ag infiltration into porous Ti6Al4V compacts. Microstructure was observed by SEM and mechanical strength was evaluated by compression tests. Results showed that the pore properties play a crucial role in the infiltration timing and the distribution of the Ag's liquid. In particular, large pores allowed the infiltration to start a few °C degrees earlier than samples with smaller pores. Three-dimensional images after infiltration showed that most of the pores were filled and the remaining ones were isolated. The resulting microstructure was composed of Ti_2Ag, α-Ti and Ag phases, indicating that the Ag diffusion occurred. Furthermore, the mechanical strength depends on the interparticle neck sizes and the Ag improves the plastic deformation reached during compression tests. The best results were obtained for the samples with larger pore sizes because the resulting mechanical properties (E = 23 GPa and σ_y = 403 MPa) are close to that of human bones, making it the best candidate as an antibacterial material for biomedical use.

Keywords: liquid phase sintering; pressureless infiltration; biomedical materials; permeability; computed microtomography; porous compacts

Academic Editors: Yuyuan Zhao and Huiping Tang

Received: 19 January 2025
Revised: 14 February 2025
Accepted: 17 February 2025
Published: 21 February 2025

Citation: Villa-Tapia, J.I.; Vergara-Hernández, H.J.; Olmos, L.; Arteaga, D.; Téllez-Martínez, J.S.; Solorio-García, V.M.; Mihalcea, E. Investigation of Pore Size Effect on the Infiltration Process of Ti6Al4V/xAg Metal Matrix Composites. *Materials* **2025**, *18*, 939. https://doi.org/10.3390/ma18050939

1. Introduction

Throughout the years, research on the use of metallic alloys in biomedical implants has been of great relevance, since these must comply with a certain series of characteristics and very specific parameters to be considered optimal for biomedical use. Different requirements are needed to use materials for long term orthopedic implants [1–3]. Among the most important are those related to the mechanical properties, which should have the most similar values possible to that of human bones. The main properties are low stiffness, high strength, and high hardness. In addition, those materials have to be biocompatible with the human body in order to avoid toxic components for the human body. In addition, those materials should demonstrate good corrosion behavior under biological fluids and, for materials used in bone joints, a wear resistance is required. In addition, it is highly recommended that materials pose a good antibacterial activity to avoid infections during the surgery process [4]. Finally, it is expected that materials also show a good osseointegration; this is reached by controlling the pore features to mimic the bone structure [5].

Currently, Ti alloys have had a great scientific development because they fulfill most of the medical requirements. These are widely used in different types of implants such as dental, external and craniofacial prostheses, and their use extends to surgical devices. The adaptation of Ti alloys for biomedical use is attributed to their high range of biocompatibility, which is the ability of a material to adapt to the human body without causing any adverse effects; this is in addition to generating an optimal cellular response so that the material in question is functional for the user. The Ti6Al4V alloy, composed of titanium (Ti), aluminum (Al) at 6% and vanadium (V) at 4%, meets the biocompatibility ranges necessary to be an applicable material for biomedical use based on the pH effects on it, as well as offering a unique combination of mechanical strength, light weight and excellent corrosion resistance [6,7].

Nevertheless, Ti6Al4V has the disadvantage of having very low antibacterial resistance, which directly affects its use as a biomedical material. It can lead to complications during and after the surgery due to bacterial infections which, directly compromises the useful life of the implant and the longevity of the human bone to which the implant is anchored [8]. In addition to this, most of the Ti6Al4V alloy parts for biomedical implants are manufactured with the sintering process due to the complex morphologies used, which results in parts with porosity and low density, also affecting their mechanical properties in the long term [9].

To overcome the issues associated with alloys, combinations with materials that have antibacterial activity, such as Ag, have been used [10–13]. It has been proven that TI6Al4V and Ag generate good antibacterial levels; M. Chen et al. [14] made the addition of Ag in different wt.% and found a higher bactericidal rate in the samples with higher wt.% of Ag added. This resulted in a considerable improvement in the alloys based on these elements for biomedical use. Different research with Ag addition [15] demonstrated a bactericidal rate around 99.99% for bacteria such as S. aureus and E. Coli, demonstrating that bacterial cell walls are completely altered by Ag.

Powder metallurgy is commonly used as a manufacturing process for biomedical prostheses because the powder sintering process has a lower production cost [16]. However, since it has a porous part, there is a risk of failure due to the mechanical behavior of these structures [17]. The addition of elements in metal matrix composites can be performed to reinforce them in different ways; an example is the assisted liquid infiltration which implies important improvements in the mechanical properties. Nevertheless, this technique has certain disadvantages such as the induction of microcracks which are detrimental to the composites [18]. On the other hand, with pressureless liquid infiltration, a satisfactory infiltration by capillary phenomena can be performed, obtaining good results without these structural defects since no external force is required to infiltrate the reinforcement material into the composite [19].

The principle of the infiltration technique is to pass some material through a porous medium in order to fill the pores [20]. Important aspects must be considered to have good infiltration levels such as wettability and permeability, in addition to inconsistencies derived from errors that may exist in the control of atmospheres and preparation of substrates that can lead to discrepancies in the contact angles. Sobczak et al. [21] concluded that it is necessary to have a control or knowledge of the surface and the atmosphere used to avoid those problems. Infiltration can be used as a method of densification of a porous material completely changing its properties. Kumar et al. [22] infiltrated a YBCO superconductor with Ag obtaining good levels of densification changing the pore volume fraction of the worked parts, in addition to their mechanical and microstructural properties. Arivazhagan et al. [23] made Mg infiltration without pressure into a porous Ti6Al4V compact, completely improving the maximum compressive strength in the infiltrated parts. The mechanical strength was improved, going from 178 MPa of the porous structures to 415 MPa of the infiltrated compacts, thus improving their mechanical properties and densification. For the infiltration into a porous compact to take place optimally, it depends on different variables—one of the most important being the existing permeability. Olmos et al. [24] carried out permeability studies in a Ti-6Al-4V porous compact through computerized microtomography, finding that the permeability is directly influenced by the pore volume fraction, the size of the channel, as well as the number and shape of the pores.

The Ti6Al4V/Ag combination, besides having good antibacterial properties, shows good results in terms of mechanical behavior. Solorio et al. [25] performed sintering of Ti6Al4V powders with different vol.% of Ag, in order to perform compression tests on the sintered compacts, obtaining better results with higher vol.% of Ag added, due to the fact that they presented densification values around 97%. The results of the combination of these materials are beneficial since their mechanical and antibacterial attributes are improved, allowing the development of a composite material with a specific microstructure which allows its use in biomedical implants [26]. The wettability of the Ag on the Ti or Ti alloys has not been studied from the wetting angle; nevertheless, it was reported that titanium is immiscible with different metals [26] as Ag [27]. In where a liquid Ag form during sintering enhanced the densification of composites. In both cases, the wettability of the alloy increased, which positively influenced the biological response of the implant.

As demonstrated, the Ti6Al4V/Ag combination is suitable for use in biomedical implants; however, it has a weakness in the density levels that it presents once sintered, which is a serious problem that compromises the life of a prosthesis. This can be solved with techniques such as gravity infiltration, thus being able to improve the mechanical properties without neglecting the antibacterial part. Since this option has not yet been studied, the selection of this technique allows it to be developed in its entirety. Therefore, the main objective of this work is to analyze the infiltration process from the point of view of the void space to be filled by liquid. Thus, porous compacts with different pore characteristics were prepared to evaluate the flow throughout the porosity with the aim of fabricating Ti6Al4V/Ag composites that could be used as bone implants. Additionally, a microstructural characterization and mechanical properties were evaluated in order to determine the viability for use them as orthopedic implants.

2. Materials and Methods

2.1. Raw Materials and Porous Compacts Fabrication

Atomized powders with spherical shapes of the Ti6Al4V alloy, produced by Raymor, Boisbriand, Canada, were sieved by passing them through meshes of different sizes, and different particle sizes were obtained: 20–45 μm, 45–75 μm and 75–106 μm. The powders obtained from the sieving process are shown in Figure 1a–c, respectively, these have different particle size percentages within the previously established ranges. Once the powders were separated, each particle size was mixed with PVA (polyvinyl alcohol) as a binder in 1% by weight solution, and then poured into a cylindrical stainless-steel die with a diameter of 6 mm and compacted at 450 MPa in an Instron 1195 mechanical testing machine to obtain green compacts. The green compacts were put through the sintering process, in which a different thermal cycle was given. Compacts with a particle size of 20–45 μm were sintered at a temperature of 1100 °C, while compacts of 45–75 μm and

75–106 μm were sintered at 1200 °C. A heating rate of 20 °C/min and the dwell time was 1 h for all samples. The sintering temperatures were chosen because a full interconnected porosity is desired to achieve a good infiltration of Ag liquid. Before sintering, PVA was removed at 500 °C for 45 min. The whole thermal cycle was carried out under atmosphere of high purity argon in a Linseis L75V vertical dilatometer, See Figure 2. To determine the density of each sintered compact, precise measurements of its mass and dimensions were performed.

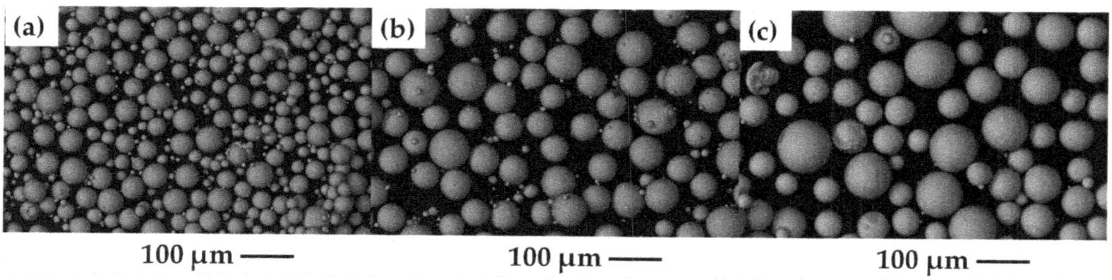

Figure 1. Initial powders Ti6Al4V, (**a**) 20–45 μm, (**b**) 45–75 μm and (**c**) 75–106 μm.

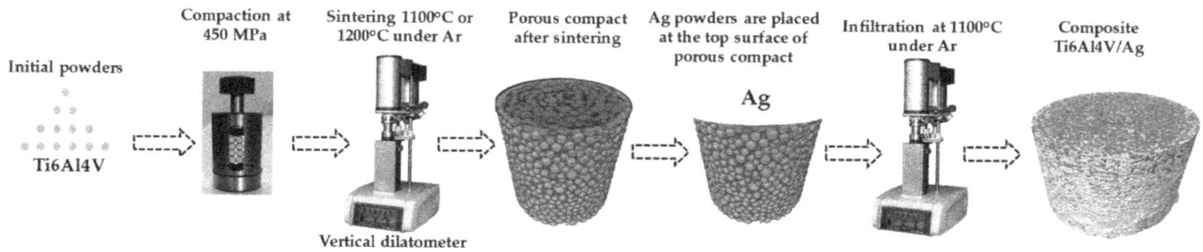

Figure 2. Schema of the experimental set up.

2.2. *Characterization of Porous Compacts*

The sintered compacts were cut into pieces of 2 mm by 2 mm on a Struers Accutom-2 precision cutter (Struers S.A.S., Champigny sur Marne cedex, France) and then microtomography was performed in a Versa 510, Zeiss nanotomograph (Carl-Zeiss-Strasse, Oberkochen, Germany), in which separate sessions were performed for each of the compacts with different particle sizes, obtaining a file with all the projections for each compact. In the image analysis, the microtomography images were converted into binary images through filters, thresholding and segmentation processing [28]. This allowed us to obtain the formation of the interparticle necks generated in the compacts. The software calculated the surfaces of the lines that divide the particles in the image, to adjust the contrast of the image itself, thus reducing the number of errors that it could contain. The analysis of the resulting microstructure in each of the compacts was obtained from 3D images with a voxel size of 3.5 μm, being worked in Avizo 2019.1® software, in addition to performing permeability simulations; these operate with binary images where the pores are represented with an intensity of 1 and Ti6Al4V with an intensity of 0. The simulation model worked in Avizo® solves the Navier–Stokes equations for laminar flow and is governed by Darcy's Law for absolute permeability. The boundary conditions that were set to the software are the inlet and outlet pressure being 130 and 100 kPa, respectively, and the viscosity of the fluid with a value of 0.00327 Pa.s, which in this case was Ag at 1100 °C, being the actual process temperature [29].

2.3. Infiltration Process and Composite Characterization

The infiltration of Ag in the sintered compacts was carried out again in the vertical dilatometer where the sintering was performed. Figure 2 illustrates the experimental steps followed to fabricate the composites by infiltration. The aim to perform the infiltration in the dilatometer was to measure the temperature at which the liquid is infiltrated into the porous compact by in situ following the axial displacement. The infiltration set up was based on having the porous Ti6Al4V compact inside a zirconium crucible of 10 mm diameter in the lower part and in the upper part of the compact the silver compacted in the form of a pellet, in addition to this an alumina rod of 1 mm in diameter and 2 cm high. All the compacts followed the same thermal cycle, being taken to a temperature of 1100 °C with a dwell time of 5 min and a heating rate of 25 °C/min. The compacts infiltrated with silver were again characterized by microtomography to perform image analysis and observe the silver infiltrated in each of the compacts. In addition, the microstructure of the infiltrated compacts was observed by scanning electron microscopy, where all the compacts were mounted on Bakelite for better handling due to the size of the pieces and prepared by roughing with SiC abrasive sandpaper starting at sandpaper 320 and ending at 2000, to be polished and reach the mirror surface using 1 μm alumina as an abrasive medium. The pieces were observed under a scanning electron microscope at 20 kV, in order to analyze the phases obtained. Compacts 6 mm wide and 5 mm high, sintered and infiltrated with silver under the same conditions as the cut compacts were subjected to mechanical compression tests in order to perform a strength evaluation. The test consisted of placing the compacts on the base of the testing machine previously used for compacting powders and then applying compression pressure until the fracture point was reached.

3. Results and Discussion

3.1. Sintering

Compacts with different particle sizes were sintered at different temperatures in order to obtain different levels of densification, as it was reported for solid state sintering of Ti6AlV powders with different particle sizes [30]. The porous structures obtained in the sintering process are shown for each particle size in Figure 3a–c, observing a structure consisting of Ti6Al4V particles and pores. It can be seen that the size of the powders used influences the volume fraction of pores obtained, which was expected due to the sintering temperatures used. For the compact fabricated with the particle size distribution of 20–45 μm, the lowest volume fraction was obtained because the contact area at the neck level is greater, which results in a higher densification, reaching a relative density of 82.5%. Samples with a particle size distribution of 45–75 μm reached a relative density of 78.7% meanwhile the relative density of samples fabricated with powders of 75–106 m reached a relative density of 70.3%. The relationship between the particle size and the densification is direct since it is observed that having a larger particle size results in a lower densification and; therefore, a higher volume fraction of pores is obtained, due to the fact that it has a smaller contact area at the interparticle necks [31].

From 3D images of gray level acquired by computed microtomography, binary images were obtained by thresholding, in order to be able to work on the whole 3D image analysis [32,33]. Figure 4a–f, shows the 3D renderings of the pre-sintered compacts, in where the Ti6Al4V and the porosity can be qualitatively observed [34]. The 75–106 μm compact was the one that presented the highest volume fraction of pores with 29.6% as shown in Table 1, denoting the previously mentioned. All the compacts presented a pore volume fraction between 15 and 30 Figure 4a–f, obtaining different levels of densification to perform the infiltration of the Ag.

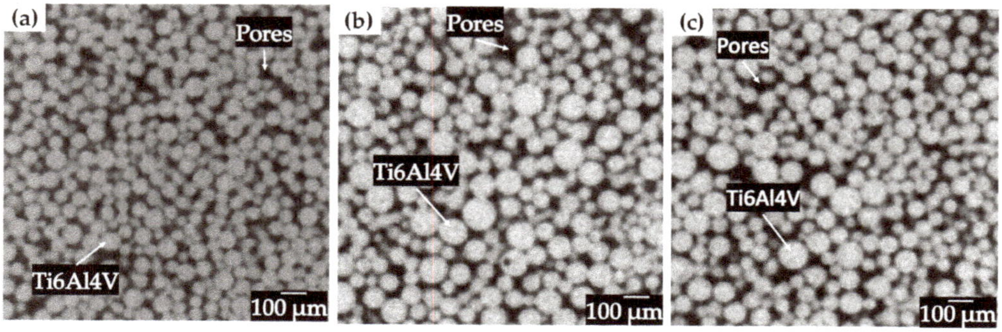

Figure 3. 2D virtual slices of Ti6Al4V sintered (**a**) 20–45 µm 1100 °C, (**b**) 45–75 µm 1200 °C and (**c**) 75–106 µm 1200 °C.

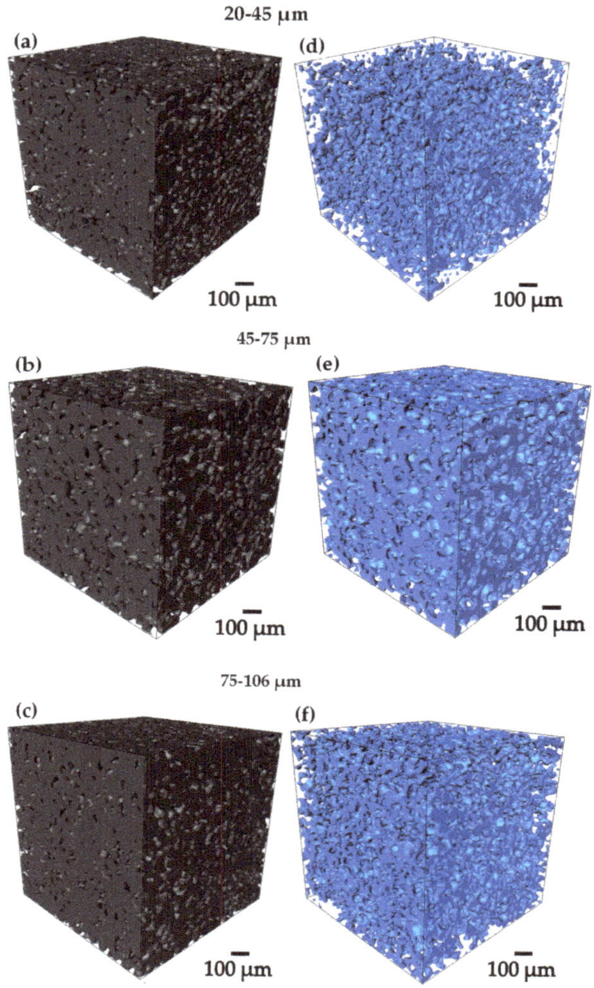

Figure 4. 3D rendering of the sintered compacts showing 2 different phases (**a–c**) show the Ti6Al4V distribution and (**d–f**) show the pore distribution obtained.

Table 1. Volume fraction of different phases of the porous compacts and composites after infiltration process.

Sample	Vol. fraction of Ti6AlV (%)	Vol. Fraction of Pores (%)	Vol. Fraction of Ag (%)	Fraction of Pores Filled with Ag (%)
		Before infiltration		
20–45 μm	82.59	17.41	--	--
45–75 μm	78.71	21.29	--	--
75–106 μm	70.38	29.62	--	--
		After infiltration		
20–45 μm	82.59	1.69	15.60	89.6
45–75 μm	78.71	1.24	20.04	94.13
75–106 μm	70.38	1.70	27.91	94.23

The segmented images were used to obtain the pore size distribution and the image analysis was performed in which the data obtained in the images were converted into tangible numerical data and interpreted in granulometry curves, shown in Figure 5. It can be seen that each compact has a different pore size distribution, which implies that the particle size used to fabricate the compacts directly influences the formation of pores, since they all followed low-temperature sintering conditions for each particle size used. The compact that presented a higher volume fraction with respect to the large pore size was the 75–106 μm compact, having pores with a size greater than 100 μm, while the 20–45 μm compact presented a higher number of pores with small sizes, from 15 μm to 40 μm. However, all the compacts showed the pattern of presenting the highest volume fraction with pore sizes from 20 μm to 60 μm. In a previous study, densification was linked to pore size as a function of the particle size used and attributed to sintering time and temperature [35].

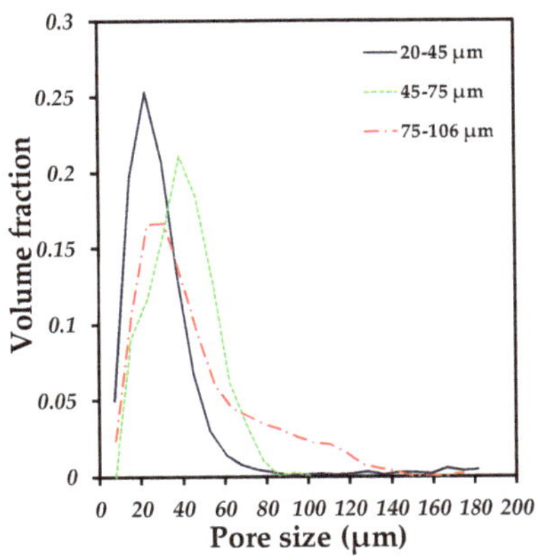

Figure 5. Pore size distribution of porous compacts fabricated with different particle size of Ti6AlV powders.

In Table 2, all values of d_{10}, d_{50} and d_{90} are given in addition to the span obtained for each compact, the d_{50} is the median pore size below which 50% of the fact that not all sample components have the same size. The d_{90} is the pore size below which 90% of the pores are contained and similarly d_{10}.

Table 2. Cumulative values of sintered compacts of Ti6Al4V.

Particle Size (μm)	d_{10} (μm)	d_{50} (μm)	d_{90} (μm)	Span
20–45	9.4920	22.6984	44.5297	1.5436
45–75	16.4927	36.6467	56.7814	1.0993
75–106	13.5212	33.8838	87.8704	2.1942

It is observed that the values are very close to each other for all the compacts; the biggest difference is in d_{90} from 75 to 106 μm, since there is a difference of approximately 30 μm. This is because for these values, there is already a greater number of data contained in this calculation. When analyzing the span, it shows that the amplitude goes according to what was obtained in the accumulated fraction curves, the pore sizes obtained follow the trend of the particle size, this in order of all the particle sizes used. The difference is mostly reflected when going from the compact with a smaller particle size (20–45 μm) to the compact with a larger size (75–106 μm).

3.2. Permeability Analysis

The simulation determines the percentage of the total volume of the compact that is occupied by pores [36,37]. These pores will be filled with Ag or air if they are pores remaining after the infiltration process. If only the volume of the pores that are interconnected is considered, the software classifies it as effective porosity and the simulation flow runs through those spaces. The fluid velocity in the pore compact is proportional to the hydraulic gradient through the permeability coefficient, thus the software does not consider it as an intrinsic property and assigns velocity units to it. The flow lines are governed by a color system indicating the velocity at which the fluid passes through the pores; the red lines indicate where the fluid velocity is faster and the blue lines indicate where velocity is slowest.

Figure 6a–c shows the flow lines through the porous sample obtained from the permeability simulations for each compact worked at 1100 °C, which was the actual temperature at which the Ag infiltration was performed, showing the difference in viscosity behavior with respect to the interaction of the liquid/solid interface between Ag and Ti6Al4V. The 20–45 μm compact, that have a greater number of smaller pores (10 μm to 60 μm) with 25% of volume fraction. It is found that the liquid Ag infiltrates the compact slowly, since despite having a good level of interconnectivity, the reduced spaces of the channels make it more difficult for the liquid Ag to pass through, which can be seen by the fact that there are many blue flow lines and very few red ones. In the 45–75 μm compact, quite a drastic change is noticed, since in this compact there is a higher concentration of pores with sizes from 30 μm to 80 μm, reaching a volume fraction of 21%. However, in this compact, there is a small fraction of 0.003 of pores with size 100 μm, besides the interconnectivity of the pores occurred in a higher percentage and the tortuosity in the channels formed was reduced. The flow lines obtained are mostly red and green, which indicates a higher velocity and a reduction in the level of tortuosity, respectively. Nevertheless, there are still blue lines, which is interpreted as interconnected pores that hinder the passage of the liquid Ag, even when the infiltration temperature is increased and the viscosity is reduced. For the 75–106 μm compact, a concentration similar to the 20–45 μm compact was obtained, with pore sizes ranging from 20 μm to 70 μm. However, there is a concentration of large pores ranging from 80 μm to 170 μm, with a volume fraction of 0.044, which led to a direct improvement in the permeability levels for this compact, because having larger pores makes the flow of Ag easier, since the levels of permeability are greatly reduced. This led to a direct improvement in the permeability levels for this compact, because having larger pores makes the flow of Ag easier, since the levels of tortuosity generated are reduced by a large proportion. It was found that infiltration depends directly on the quantity and size of the pores generated. The interconnectivity plays an important role in infiltration as can be seen in permeability simulations, but it is worth noting that despite obtaining greater interconnectivity in the 20–45 μm compact in comparison with the 45–75 μm compact, better infiltration results were obtained, due to the aforementioned phenomenon. For the 75–106 μm compact, the pore size and interconnectivity results were very favorable.

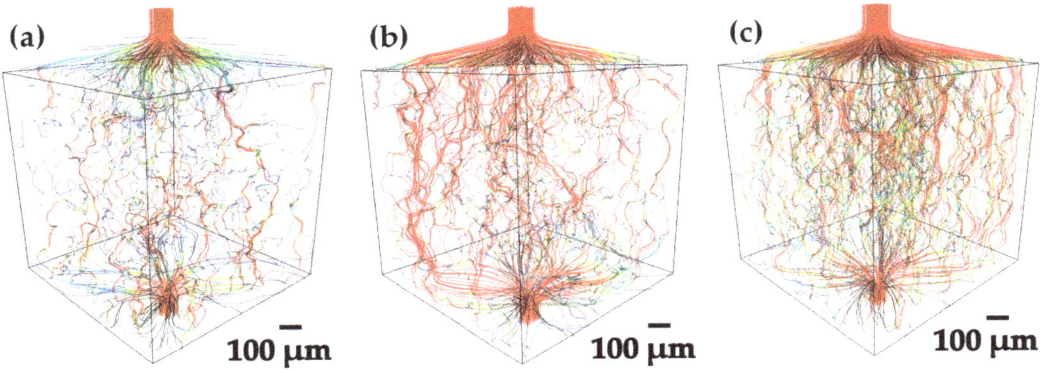

Figure 6. 3D volume rendering of permeability simulations for the flow of Ag through the pores of Ti6Al4V compacts: (**a**) 20–45 μm, (**b**) 45–75 μm and (**c**) 75–106 μm.

A 3D skeleton was made for all compacts with different particle size Figure 7a–c to ascertain whether the levels of interconnectivity were good [38], as this was required to perform the infiltration, in the generated skeleton images both lines and nodes can be observed, the lines correspond to the channels generated in the compact and are called "segments" and the "nodes" are the critical points where the lines converge and connect, generating different paths and interconnectivity between the channels [39]. Table 3 shows the values obtained for nodes and segments for each compact.

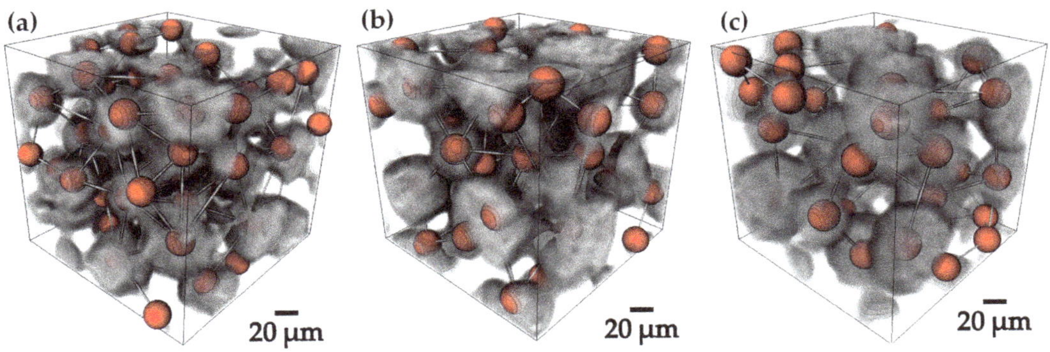

Figure 7. 3D rendered skeletons showing nodes and segments of the sintered Ti6Al4V compacts: (**a**) 20–45 μm, (**b**) 45–75 μm and (**c**) 75–106 μm.

Table 3. Segments and nodes generated in sintered Ti6Al4V compacts.

Particle Size (μm)	Segments	Nodes
20–45	16,306	11,643
45–75	18,706	21,565
75–106	21,889	37,577

The 20–45 μm compact had the lowest number of nodes and segments; however, it can be seen that, due to the particle size, the channels are of a very small size, unlike the 45–75 μm compact where the main difference lies in the size of the channels. Although there are fewer, they are completely affected by the particle size, since they are larger, and this generates a better permeability in the compact. For the 75–106 μm compact, 21,889 nodes and 37,577 segments were obtained, being the compact with the highest number of formed and interconnected channels, in addition to having a good percentage of channels with a large size.

3.3. Dilatometry Analysis

The displacement was directly plotted as a function of temperature, allowing us to know the exact temperature point at which the Ag infiltrates in each compact and the range in which the Ag infiltrates completely in all the pores. Figure 8 shows the dilatometry curves for each compact where the infiltration range for all the compacts is affected between 950 °C and 970 °C. The melting of the Ag is reflected immediately upon reaching its theoretical melting point at 950 °C. However, the displacement is still affected as a function of the temperature increase, since the equipment takes the parameters of how the porosity of the compact is being filled by the Ag. All the compacts infiltrated at different temperature ranges; the compact that infiltrated first was the 75–106 μm compact, having its critical infiltration point at 953.6 °C (this is understood as the Ag occupied all the available empty spaces). Nonetheless, the displacement activity started from 950 °C—this completely governed as a function of the porosity and tortuosity generated.

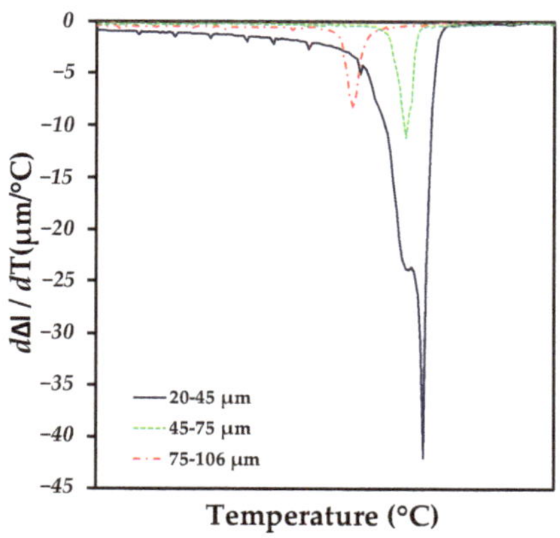

Figure 8. Infiltration curves of Ti6Al4V/xAg compacts showing exact temperature points where pores are filled with Ag.

Wettability is a factor that can be discarded, since, as mentioned by Assael M.J. et al., the viscosity of Ag does not change much as a function of the temperature range used, which was previously proven in permeability simulations [29]. The 45–75 μm compact infiltrated at 960.5 °C and its range of displacement activity started at 956.8 °C. On the other hand, the 20–45 μm compact was the one most affected by displacement, it was found that this one had two critical points of infiltration starting to infiltrate at 960.9 °C and finishing infiltration at 962. 9 °C, due to the fact that the amount of pores with a small size was seen in greater proportion in this compact once sintered. In addition to the tortuosity obtained with a large number of nodes generated but with channels of reduced size, making it difficult for the Ag to fill the pores in their entirety excluding the isolated pores.

3.4. Ag Infiltration

The way of working the new phases in the analysis of the microtomographies changes, since the contrasts of the initial image made it necessary to adjust the thresholds of the images and thus obtain a binary image where the remaining pores could be visualized, the threshold adjustment levels of the images depended on each compact worked, since each one had a different number of voxels [40]. However, the results in volume fraction directly agree with the data taken before performing the Ag infiltration, taking into account a 100% pore volume fraction since the beginning of the analysis.

Ag infiltration was performed on Ti6Al4V compacts and 3D images of the infiltrated compacts were acquired by microtomography, as shown in Figure 9a–c. The difference in densities is reflected in the colors recorded in the images, the Ti6Al4V particles are recorded in dark gray, while the infiltrated Ag was recorded in light gray and the remaining pores are completely black. It is found that the infiltration process reached to fill most of the interconnected pores and leaving empty the remaining pores, since being completely isolated, they are not reached by the Ag and are completely empty. The image analysis process followed the same processing route that was used for the images of the pre-sintered compacts, the difference was in the Ag phase and how the images were processed with this new variable.

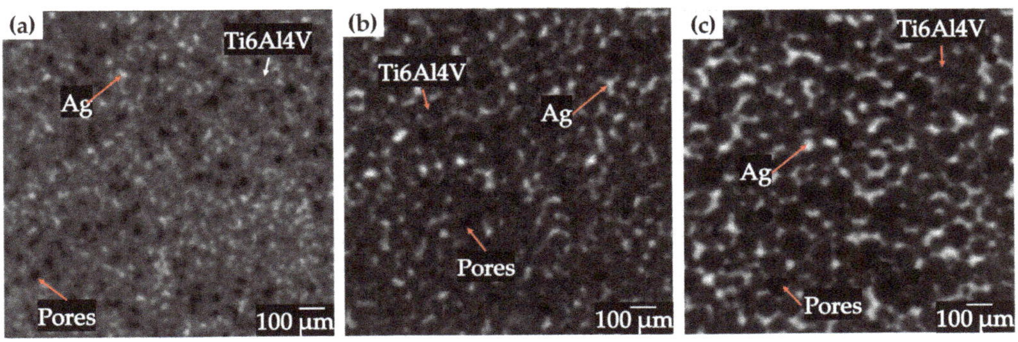

Figure 9. 2D virtual slices of Ti6Al4V/xAg infiltrated: (**a**) 20–45 μm, (**b**) 45–75 μm and (**c**) 75–106 μm.

The compact with the highest volume fraction of Ag infiltrated is the 75–106 μm compact with 27.91% infiltrated. This due to the remaining porosity after sintering as discussed above in the 3D image analysis of the pre-sintered compacts. Numerical flow simulations and permeability analysis also agreed that this porous compact presented the best conditions for the infiltration process. The 20–45 μm and 45–75 μm compacts presented a lower volume fraction of Ag infiltrated, no more than 21% as shown in Table 1. However, they are still good infiltration levels to work with. All the compacts had a remaining porosity of less than 2% in volume, which indicates that the conditions worked in the pre-sintering process were optimal because they allowed to generate a good interconnectivity and tortuosity in the compacts [41,42].

In Figure 10a–f the 3D renderings of the Ag infiltrated in the compacts and the remaining porosity obtained after the process can be observed qualitatively. However, the results in volume fraction agrees directly with the data taken before performing the Ag infiltration, taking into account a 100% in volume from the beginning of the analysis.

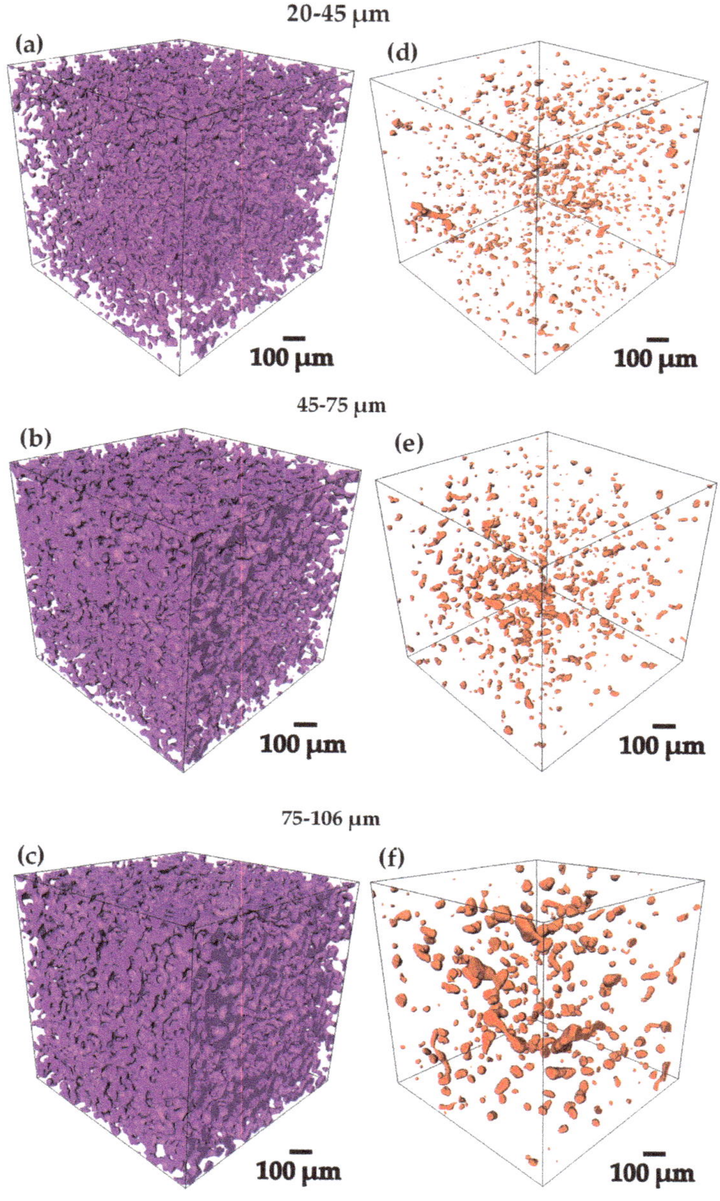

Figure 10. 3D rendering of the infiltrated compacts showing 2 different phases (**a–c**) show the Ag distribution and (**d–f**) show the remaining pores distribution obtained.

Figure 11 visualizes the microstructure in each compact after the infiltration process, for all compacts the same microstructure was obtained, composed by the phases: Ag, Ti_2Ag and α-Ti. The Ti2Ag and α-Ti phases were found diffused within the titanium particles in the form of lamellae and small islands. The diffusion of Ag into the solid Ti6AlV particles generetes the formation of lamellae from houndreds of nanometers to a few microns of Ti_2Ag phases, as shown in Figure 11d. The diffusion of some atoms into the cristalyne net of Ti6AlV phase induces the stabilization of α-Ti. This has been reported on Ti-Ag alloys fabricated by either powder matallurgy [14] or melting [43]. Shi et al., attributed that the antibacterial behavior is reflected by the Ti_2Ag phase although the Ag phase also have antibacterial activity due to the release of ions, therefore all the compacts are considered antibacterial [44]. There are two points of view regarding the mechanisms of antibacterial action Ag. The first and widely accepted is the one related to the released metal ions killing bacteria, in where the dissolution of silver ions may also cause free radicals on the titanium surface that destroy the bacterial structure and so generate the silver ion sterilization [45,46]. The second is related to the contact sterilization; it is considered that the presence of Ag particles on the surface of the metal can generate normal physiological metabolic disorders of bacterial cell membrane, thus, leading to bacteria death [47]. Shi et al. [44] demonstrated that the precipitation of Ti_2Ag in a contact sterilization mode plays a major role in the antibacterial ability of the Ti- Ag alloy than the silver ion release. The volume fraction of Ag obtained after the infiltration process are sufficient to have a good bactericidal rate since small amounts of Ag promote antibacterial activity [48].

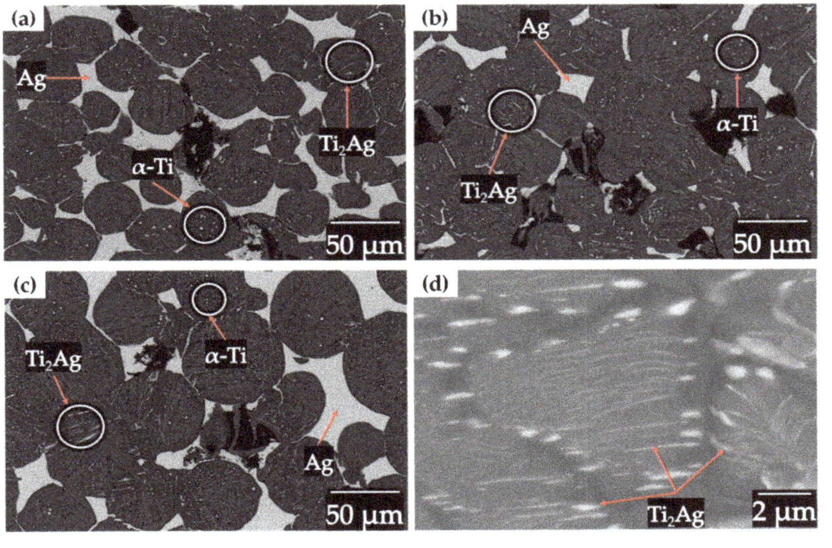

Figure 11. Backscattered electron image of the Ti6Al4V/xAg compacts: (**a**) 20–45 μm, (**b**) 45–75 μm, (**c**) 75–106 μm and (**d**) Ti_2Ag lamellae inside of Ti6Al4V particles.

3.5. Compression Tests

The infiltrated compacts were taken to the fracture point in order to know their mechanical resistance to compression. Figure 12 shows the compacts after the test and how the fracture propagated, which propagated through all the interparticle necks formed and also through the titanium particles. Figure 13a shows that the fracture propagation took place through the particles because a higher energy is required to break the necks formed, which was reflected in all the compacts [49]. It was also shown that Ag is mainly deformed during the compression tests, which brings a larger ductility to the composite (Figure 13b,c). In addition, it was observed that Ag covering the Ti6AlV particles also is break down when the strenght of the Ag reached their limit, Figure 13d. This suggests that the fracture behavior is composed of two mechanicsms, the first one is the plastic deformation of the Ag that covers the Ti6AlV particles that cans also be fractured. The second one leads to the fracture propagation once the limit of particles necks is reached, which turns in the fracture of composites.

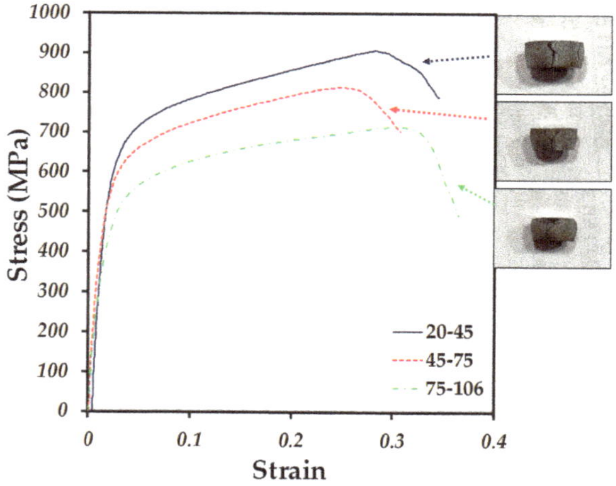

Figure 12. Stress–strain compression curves of Ti6Al4V/xAg with different particle size showing the fractures for each compact.

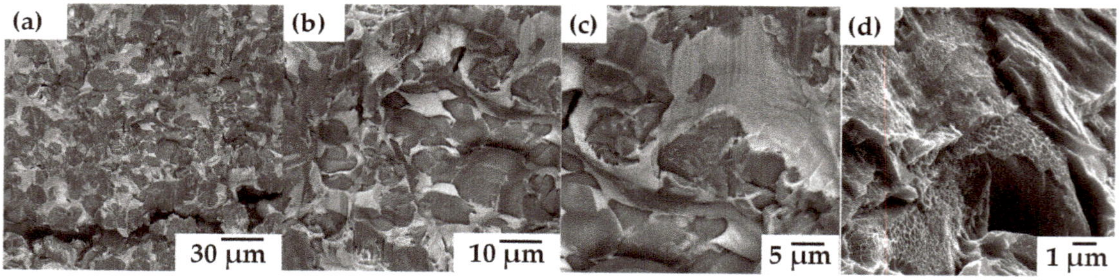

Figure 13. SEM images of fractured samples after compression tests at different magnifications; (**a**) 1000×, (**b**) 3000×, (**c**) 5000× and (**d**) 20,000×.

In the strain-stress curves obtained from the compression tests, the compact with the best mechanical resistance to compression was the 20–45 μm compact, since having a greater number of particles forming necks generates a greater mechanical resistance.

Table 4 shows the yield stress (σ_y) and Young's modulus (E) for each compact. The 20–45 μm compact had a yield stress of 504 MPa and a Young's modulus of 38 GPa, while the compact that had a lower mechanical resistance to compression was the 75–106 μm compact, due to the opposite case of the compact with smaller particle size, Since it has a smaller contact area between the particles, less interparticle necks are generated, which favors this behavior, resulting in a yield stress of 403 MPa and a Young's modulus of 23 GPa. It is observed that the infiltration of Ag does not greatly influence the mechanical properties because the compacts have a behavior characteristic of compacts with those particle sizes.

Table 4. Modulus of elasticity and yield stress obtained in compression tests of Ti6Al4V/xAg compacts.

Particle Size (μm)	Yield Stress σ_y (MPa)	Young's Modulus E (GPa)
20–45	504	38
45–75	463	30
75–106	403	23

The 75–106 μm compact, despite having a lower mechanical strength, is the option that would be the most viable for use in a biomedical prosthesis, since its mechanical properties are the most similar to those of human bones. Donald T. et al. reported values of Young's modulus (E) for human bones in a range of 15 to 25 MPa, depending on the age, diet and physical routine of the patient [50]. Thus, the 75–106 μm compact falls within the range of mechanical properties of real bone; therefore, at the time of anchoring between the human bone and the compact of these characteristics there will be no mechanical decompensation, thus ensuring the life of the compact and the longevity of the user's bone, avoiding compromising both parts [51].

4. Conclusions

The pressureless gravity infiltration of Ag into porous Ti6Al4V solid compacts with different particle sizes sintered at low-temperature has been successfully performed. It has been demonstrated that the larger the pore size, the better the filling in the gravity infiltration process, as the higher the liquid velocity, the better the permeability of the compact. In addition, the diffusion of Ag in the Ti6Al4V particles leads to the formation of the Ti2Ag phase in the form of lamellae.

The 3D analysis by microtomography made it possible to determine the pores' features through which the liquid silver passes during the infiltration process. The best infiltration results were obtained in the compacts with a particle size of 75–106 μm, since they allowed a higher densification and a faster filling of the compact, which benefited the mechanical properties of this compact to be the most similar to those of human bones.

Author Contributions: Conceptualization, L.O.; Data curation, J.S.T.-M.; Formal analysis, L.O.; Investigation, J.I.V.-T., H.J.V.-H. and V.M.S.-G.; Methodology, J.I.V.-T., H.J.V.-H. and V.M.S.-G.; Project administration, H.J.V.-H.; Software, D.A.; Supervision, D.A., J.S.T.-M. and E.M.; Validation, D.A.; Writing—original draft, J.I.V.-T. and L.O.; Writing—review & editing, E.M. All authors have read and agreed to the published version of the manuscript.

Funding: This research was funded by the "Coordination de la Investigación Científica" of the Michoacana University and by the National Council for Science and Technology CONAHCYT via Master Scholarship No. 815859.

Institutional Review Board Statement: Not applicable.

Informed Consent Statement: Not applicable.

Data Availability Statement: The data presented in this study are available on request from the corresponding author. The data are not publicly available due to the fact that are part of an ongoing study.

Acknowledgments: The authors thank to the Tecnológico Nacional de México TecNM/ITMorelia for the financial support and the facilities provided for the development of this study. We would also like to thank the Laboratory "LUMIR" of Instituto de Geociencias, UNAM, Juriquilla for the 3D image acquisition and processing.

Conflicts of Interest: The authors declare that they have no conflict of interest.

References

1. Attar, H.; Ehtemam-Haghighi, S.; Soro, N.; Kent, D.; Dargusch, M.S. Additive manufacturing of low-cost porous titanium-based composites for biomedical applications: Advantages, challenges and opinion for future development. *J. Alloys Compd.* **2020**, *827*, 154263. [CrossRef]
2. Geetha, M.; Singh, A.K.; Asokamani, R.; Gogia, A.K. Ti based biomaterials, the ultimate choice for orthopaedic implants–a review. *Prog. Mater. Sci.* **2009**, *54*, 397–425. [CrossRef]
3. Abd-Elaziem, W.; Darwish, M.A.; Hamada, A.; Daoush, W.M. Titanium-Based alloys and composites for orthopedic implants Applications: A comprehensive review. *Mater. Des.* **2024**, *241*, 112850. [CrossRef]
4. Jiao, J.; Zhang, S.; Qu, X.; Yue, B. Recent advances in research on antibacterial metals and alloys as implant materials. *Front. Cell. Infect. Microbiol.* **2021**, *11*, 693939. [CrossRef]
5. Kreve, S.; Ferreira, I.; da Costa Valente, M.L.; Dos Reis, A.C. Relationship between dental implant macro-design and osseointegration: A systematic review. *Oral Maxillofac. Surg.* **2024**, *28*, 1–14. [CrossRef]
6. Rack, H.; Qazi, J. Titanium alloys for biomedical applications. *Mater. Sci. Eng. C* **2006**, *26*, 1269–1277. [CrossRef]
7. Gil, F.J.; Planell, J.A. Aplicaciones biomédicas del titanio y sus aleaciones. *Soc. Ibérica Biomecánica Biomater.* **1993**, *1*, 34–42. [CrossRef]
8. Li, M.; Zheng, M.; Zhu, Y.; Xia, H.; Yao, M.; Chu, X.; Wang, X.; Yang, K. Toward a Molecular Understanding of the Antibacterial Mechanism of Copper-Bearing Titanium Alloys against Staphylococcus aureus. *Adv. Health Mater.* **2016**, *5*, 557–566. [CrossRef]
9. Yan, Y.; Nash, G.L.; Nash, P. Effect of density and pore morphology on fatigue properties of sintered Ti-6Al-4V. *Int. J. Fatigue* **2013**, *55*, 81–91. [CrossRef]
10. Hassanien, A.S.; Khatoon, U.T. Synthesis and characterization of stable silver nanoparticles, Ag-NPs: Discussion on the applications of Ag-NPs as antimicrobial agents. *Phys. B Condens. Matter* **2019**, *554*, 21–30. [CrossRef]
11. Liao, K.H.; Ou, K.L.; Cheng, H.C.; Lin, C.T.; Peng, P.W. Effect of silver on antibacterial properties of stainless steel. *Appl. Surf. Sci.* **2010**, *256*, 3642–3646. [CrossRef]
12. Melaiye, A.; Youngs, W.J. Silver and its application as an antimicrobial agent. *Expert Opin. Ther. Pat.* **2005**, *15*, 125–130. [CrossRef]
13. Chernousova, S.; Epple, M. Silver as antibacterial agent: Ion, nanoparticle, and metal. *Angew. Chem.-Int. Ed.* **2013**, *52*, 1636–1653. [CrossRef] [PubMed]
14. Chen, M.; Zhang, E.; Zhang, L. Microstructure, mechanical properties, bio-corrosion properties and antibacterial properties of Ti-Ag sintered alloys. *Mater. Sci. Eng.* **2016**, *62*, 350–360. [CrossRef]
15. Cho, K.H.; Park, J.E.; Osaka, T.; Park, S.G. The study of antimicrobial activity and preservative effects of nanosilver ingredient. *Electrochim. Acta* **2005**, *51*, 956–960. [CrossRef]
16. Cantero, S.M.; Oldani, C.R.; Lucci, R.O. Desarrollo de Estructuras Porosas de Titanio para Implantes Biomédicos Obtenidas por Gel-casting y Sinterización. *AJEA* **2019**, *04*, 25–31. [CrossRef]
17. Holmes, J.; Queeney, R.A. Fatigue crack initiation in a porous steel. *Powder Metall.* **1985**, *28*, 231–235. [CrossRef]
18. Sha, J.; Lu, Z.; Sha, R.; Zu, Y.; Dai, J.; Xian, Y.; Zhang, W.; Cui, D.; Ya, C. Improved wettability and mechanical properties of metal coated carbon fiber-reinforced aluminum matrix composites by squeeze melt infiltration technique. *Trans. Nonferrous Met. Soc. China* **2021**, *31*, 317–330. [CrossRef]
19. Lee, K.B.; Kim, Y.S.; Kwon, H. Fabrication of Al-3 Wt pct Mg matrix composites reinforced with Al_2O_3 and SiC particulates by the pressureless infiltration technique. *Met. Mater. Trans. A* **1998**, *29*, 3087–3095. [CrossRef]
20. Hwang, W.; Redner, S. Infiltration through porous media. *Phys. Rev. E* **2001**, *63*, 021508. [CrossRef]
21. Sobczak, N.; Singh, M.; Asthana, R. High-temperature wettability measurements in metal/ceramic systems—Some methodological issues. *Curr. Opin. Solid State Mater. Sci.* **2005**, *9*, 241–253. [CrossRef]
22. Kumar, N.D.; Rajasekharan, T.; Seshubai, V. YBCO/Ag composites through a preform optimized infiltration and growth process yield high current densities. *Supercond. Sci. Technol.* **2011**, *24*, 085005. [CrossRef]
23. Arivazhagan, A.; Venugopal, P.R.; Mohammad, A.; Ravi, K.R. Influence of magnesium infiltration on compressive behavior of additively manufactured porous Ti6Al4V structure. *J. Test. Eval.* **2022**, *49*, 4326–4343. [CrossRef]
24. Olmos, L.; Bouvard, D.; Cabezas-Villa, J.L.; Lemus-Ruiz, J.; Jiménez, O.; Arteaga, D. Analysis of Compression and Permeability Behavior of Porous Ti6Al4V by Computed Microtomography. *Met. Mater. Int.* **2019**, *25*, 669–682. [CrossRef]
25. Solorio, V.M.; Vergara-Hernández, H.J.; Olmos, L.; Bouvard, D.; Chávez, J.; Jimenez, O.; Camacho, N. Effect of the Ag addition on the compressibility, sintering and properties of Ti6Al4V/xAg composites processed by powder metallurgy. *J. Alloys Compd.* **2022**, *890*, 161813. [CrossRef]
26. Nagase, T.; Matsumoto, M.; Fujii, Y. Microstructure of Ti-Ag immiscible alloys with liquid phase separation. *J. Alloys Compd.* **2018**, *738*, 440–447. [CrossRef]
27. Nagase, T. Alloy Design, Thermodynamics, and Electron Microscopy of Ternary Ti-Ag-Nb Alloy with Liquid Phase Separation. *Materials* **2020**, *13*, 5268. [CrossRef] [PubMed]

28. Brodersen, C.R.; Lee, E.F.; Choat, B.; Jansen, S.; Phillips, R.J.; Shackel, K.A.; Mcelrone, A.J.; Matthews, M.A. Automated analysis of three-dimensional xylem networks using high-resolution computed tomography. *New Phytol.* **2011**, *191*, 1168–1179. [CrossRef] [PubMed]
29. Assael, M.J.; Kalyva, A.E.; Antoniadis, K.D.; Banish, R.M.; Egry, I.; Wu, J.; Kaschnitz, E.; Wakeham, W.A. Reference data for the density and viscosity of liquid antimony, bismuth, lead, nickel and silver. *High Temp.-High Press.* **2012**, *41*, 161–184.
30. Cabezas-Villa, J.L.; Lemus-Ruiz, J.; Bouvard, D.; Jiménez, O.; Vergara-Hernández, H.J.; Olmos, L. Sintering study of Ti6Al4V powders with different particle sizes and their mechanical properties. *Int. J. Miner. Metall. Mater.* **2018**, *25*, 1389–1401. [CrossRef]
31. Esen, Z.; Tarhan-Bor, E.; Bor, Ş. Characterization of loose powder sintered porous titanium and TI6Al4V alloy. *Turk. J. Eng. Environ. Sci.* **2009**, *33*, 207–219. [CrossRef]
32. Li, Q.; Liu, D.; Cai, Y.; Zhao, B.; Qiu, Y.; Zhou, Y. Scale-span pore structure heterogeneity of high volatile bituminous coal and anthracite by FIB-SEM and X-ray μ-CT. *J. Nat. Gas Sci. Eng.* **2020**, *81*, 103443. [CrossRef]
33. Lawrence, M.; Jiang, Y. Porosity, pore size distribution, micro-structure. In *Bio-Aggregates Based Building Materials*; RILEM State-of-the-Art Reports; Springer: Dordrecht, The Netherlands, 2017; Volume 23, pp. 39–71. [CrossRef]
34. Lame, O.; Bellet, D.; Di, M.; Bouvard, D.M. In situ microtomography investigation of metal powder compacts during sintering. *Nucl. Instrum. Methods Phys. Res.* **2003**, *200*, 287–294. [CrossRef]
35. Razavi-Tousi, S.S.; Yazdani-Rad, R.; Manafi, S.A. Effect of volume fraction and particle size of alumina reinforcement on compaction and densification behavior of Al-Al_2O_3 nanocomposites. *Mater. Sci. Eng. A* **2011**, *528*, 1105–1110. [CrossRef]
36. Zhang, L.; Wenlong, J.; Yongfei, Y.; Hainan, Y.; Yaohao, G.; Hai, S.; Jianlin, Z.; Jun, Y. The investigation of permeability calculation using digital core simulation technology. *Energies* **2019**, *12*, 3273. [CrossRef]
37. Watanabe, N.; Hirano, N.; Tsuchiya, N. Determination of aperture structure and fluid flow in a rock fracture by high-resolution numerical modeling on the basis of a flow-through experiment under confining pressure. *Water Resour Res.* **2008**, *44*, W06412. [CrossRef]
38. Li, H.; Liu, Y.; Colombo, P.; Li, W.; Liu, Y.; Hu, K.; Lu, Z. The influence of sintering procedure and porosity on the properties of 3D printed alumina ceramic cores. *Ceram Int.* **2021**, *47*, 27668–27676. [CrossRef]
39. Kuentzer, N.; Simacek, P.; Advani, S.G.; Walsh, S. Permeability characterization of dual scale fibrous porous media. *Compos. Part A Appl. Sci. Manuf.* **2006**, *37*, 2057–2068. [CrossRef]
40. Li, Y.; Chi, Y.; Han, S.; Zhao, C.; Miao, Y. Pore-throat structure characterization of carbon fiber reinforced resin matrix composites: Employing Micro-CT and Avizo technique. *PLoS ONE* **2021**, *16*, e0257640. [CrossRef]
41. De-Oro-Calderon, R.; Gierl-Mayer, C.; Danninger, H. Fundamentals of Sintering: Liquid Phase Sintering. In *Encyclopedia of Materials: Metals and Alloys*; Elsevier: Amsterdam, The Netherlands, 2022; Volume 3, pp. 481–492. [CrossRef]
42. Ghanbarian, B.; Hunt, A.G.; Ewing, R.P.; Sahimi, M. Tortuosity in Porous Media: A Critical Review. *Soil Sci. Soc. Am. J.* **2013**, *77*, 1461–1477. [CrossRef]
43. Fu, S.; Zhang, Y.; Qin, G.; Zhang, E. Antibacterial effect of Ti-Ag alloy motivated by Ag-containing phases. *Mater. Sci. Eng. C* **2021**, *128*, 112266. [CrossRef]
44. Shi, A.; Zhu, C.; Fu, S.; Wang, R.; Qin, G.; Chen, D.; Zhang, E. What controls the antibacterial activity of Ti-Ag alloy, Ag ion or Ti2Ag particles? *Mater. Sci. Eng. C* **2020**, *109*, 110548. [CrossRef]
45. Marambio-Jones, C.; Hoek, E.M. A review of the antibacterial effects of silver nanomaterials and potential implications for human health and the environment. *J. Nanopart. Res.* **2010**, *12*, 1531–1551. [CrossRef]
46. Li, M.; Nan, L.; Xu, D.; Ren, G.; Yang, K. Antibacterial performance of a Cu-bearing stainless steel against microorganisms in tap water. *J. Mater. Sci. Technol.* **2015**, *31*, 243–251. [CrossRef]
47. Chen, M.; Yang, L.; Zhang, L.; Han, Y.; Lu, Z.; Qin, G.; Zhang, E. Effect of nano/micro-Ag compound particles on the bio-corrosion, antibacterial properties and cell biocompatibility of Ti-Ag alloys. *Mater. Sci. Eng. C* **2017**, *75*, 906–917. [CrossRef]
48. Cai, D.; Yu, Z.; Yang, L.; Nie, J.J.; Wang, R.; Chen, D.; Zhang, E. Effect of Ag content on the microstructure and antibacterial activity of biomedical titanium alloy Ti-13Nb-13Zr-xAg (x = 7,10 wt%). *Mater. Lett.* **2024**, *354*, 135389. [CrossRef]
49. Svobodat, J.; Riedel, H. New solutions describing the formation of interparticle necks in solid-state sintering. *Acta Met. Mater.* **1995**, *43*, 1–10. [CrossRef]
50. Reilly, D.T.; Burstein, A.H.; Frankel, V.H. The Elastic Modulus for Bone. *J. Biomech.* **1974**, *7*, 271–275. [CrossRef] [PubMed]
51. Güden, M.; Çelik, E.; Hizal, A.; Altindiş, M.; Çetiner, S. Effects of compaction pressure and particle shape on the porosity and compression mechanical properties of sintered Ti6Al4V powder compacts for hard tissue implantation. *J. Biomed. Mater. Res. B Appl. Biomater.* **2008**, *85*, 547–555. [CrossRef] [PubMed]

materials

MDPI

Article

Effects of the Space Holder Shape on the Pore Structure and Mechanical Properties of Porous Cu with a Wide Porosity Range

Jian Xiao [1,*], Yanping He [1], Wenjun Ma [1], Yiheng Yue [1] and Guibao Qiu [2,*]

1 School of Material Science and Engineering, Jiangxi University of Science and Technology, Ganzhou 341000, China; hyp1234562023@163.com (Y.H.); 15359783512@163.com (W.M.); 19142171189@163.com (Y.Y.)

2 College of Materials Science & Engineering, Chongqing University, Chongqing 400044, China

* Correspondence: xiaojian@jxust.edu.cn (J.X.); qiuguibao@cqu.edu.cn (G.Q.)

Abstract: Porous copper (Cu), with varying porosities, has been made using carbamide as a space holder through the powder metallurgy route. Two shapes of carbamide particles were used, (i) needle-like and (ii) spherical, in order to investigate the effect of the space holder shape on the pore structure and mechanical properties of porous Cu. The compressive deformation behavior of porous Cu was studied under a compression test. The pores' structural characteristics and mechanical properties of the porous Cu varied significantly with the shape of the space holder. Although the effect of the space holder shape on the porosity was not regular, the effect on the mechanical properties was regular. The stress increased monotonically with the increase in the strain, and strain hardening occurred at the plastic yield stage. The elastic modulus and yield strength followed the power law, with the relative density irrespective of the space holder shape. The empirical constants associated with different empirically developed power law relations were different, according to the shape of space holder. A quantitative relationship between the elastic modulus and yield strength and the spacer content was obtained to control the mechanical properties of the present porous Cu or other porous metals and metal foams using the well-known space holder method.

Keywords: porous metal; metal foam; copper; powder metallurgy; compression performance

Citation: Xiao, J.; He, Y.; Ma, W.; Yue, Y.; Qiu, G. Effects of the Space Holder Shape on the Pore Structure and Mechanical Properties of Porous Cu with a Wide Porosity Range. *Materials* **2024**, *17*, 3008. https://doi.org/10.3390/ma17123008

Academic Editors: Jordi Sort and Chih-Chun Hsieh

Received: 8 May 2024
Revised: 7 June 2024
Accepted: 17 June 2024
Published: 19 June 2024

1. Introduction

Porous copper (Cu) has been widely used for heat dissipation, as a battery motor and a catalyst carrier, as well as for other applications [1]. The use of space-holding filler materials is an indirect method to fabricate porous Cu [2]. In this method, the Cu powder is either filled into a "dry" bulk of fillers, a suitable solvent, or even an organic binder to mix the space holders and the metal powders.

For porous Cu based on the space holder method, the earliest literature can be traced back to 2005 [3]. The paper described a lost carbonate sintering process for manufacturing open cell metal foams. Cu foams with a porosity of 50–85% and cell sizes of 53–1500 μm were manufactured. Since then, the use of this method to prepare porous Cu with a foam structure has become increasingly popular, and the space holder is still in use today. In the past few decades, a large number of studies have focused on the effect of the content and particle size of the space holder on the pore structure and mechanical properties of porous Cu. According to these studies, the porosity of the porous Cu increased with the increase in the space holder content, while the mechanical properties decreased. For example, Shirjang et al. [4] stated that the porosity of copper foams increased from 50.2 to 71.8%, and the plateau stress decreased from 83 ± 3.1 MPa to 20 ± 1.8 MPa, as the content of K_2CO_3 used as the space holder increased from 50% to 70%. The porosity of the porous copper increased from 38.48 ± 0.91% to 66.47 ± 1.06%, and the yield strength decreased from 42.8 ± 4.6 MPa to 6.9 ± 0.9 MPa when the space holder of bimodal K_2CO_3: NaCl was 1:2, as their contents increased from 30% to 70%, according to Ray et al. [5] The total porosity of porous copper increased from 27.5% to 67.1%, and the elastic modulus (C_{11}) decreased from 59.4 ± 0.2 GPa to 1.0 ± 0.2 GPa when the content of K_2CO_3 used as the space holder increased from 0 to 54.4%, according to Jana et al. [6]. Luo et al. [7] showed that the elastic modulus of titanium foam decreased with the increasing size of the magnesium particles, while the porosity decreased, etc. However, little attention was paid to the shape of the space holder, even when the study of the shape of the metal powder extended the scope to the entire porous metal. For example, Jain et al. [8] studied the effects of the particle shape on the microstructure and compressive response of 316L SS foam using the space holder technique. Mondal et al. [9] studied the effect of the particle shape and strain rate on the microstructure and compressive deformation response of pure Ti-foam made using acrowax as the space holder. The existing studies showed that the particle size of the space holder had little influence on the pore structure and mechanical properties of porous Cu compared with its content. However, it is not clear whether the influence of the shape of space holder on the pore structure and mechanical properties of porous Cu is negligible compared with the content.

Therefore, the effect of the space holder shape relative to its content on the pore structure and mechanical properties of porous Cu was studied in this work. Based on the author's previous research [10–12], spherical carbamide with two sizes was selected as the space holder due to its high quality, low price, and easy removal at low temperature.

2. Materials and Methods

The raw material of the metal powder was selected as Cu powder (purity: ~99.9%, size: ≤50 μm) with a coralloid shape (see Figure 1a), using an electrolytic process. The Cu powder was purchased from Beijing Hongyu New Material Technology Co., Ltd. (Beijing, China). The carbamide with two shapes, needlelike and spherical, was used as the space holder, as shown in Figure 1b,c. The carbamide was purchased from Xilong Scientific Co., Ltd. (Chengdu, China) and sieved with 40–80 mesh for the needlelike particles, while the spherical carbamide was 0.85–3 mm. The diameter and height of the sample were designed as 20 mm and 8 mm, respectively. The volume content of the space holder was set as 10, 20, 30, 40, 50, 60, 70, and 80 vol.%, respectively. According to the density of the Cu powder (8.96 g/cm^3) and carbamide (1.335 g/cm^3), the weight of the Cu powder and carbamide particles were calculated. The weights of the metal powder and needlelike space holder particles were increased by 0.05 g, considering their mass loss. For spherical carbamide, it was formulated according to the calculated value due to its large particle size.

We poured the measured Cu powder and carbamide into the mortar, added an appropriate amount of anhydrous ethanol, and stirred. After mixing evenly, it was put into the steel mold for pressing and forming, and the pressing pressure and holding time were 200 MPa and 0.5 min, respectively. A layer metal mesh support in a porcelain boat was used to place the obtained green compacts. One step of heat treatment (see Figure 1d was used to determine the compacts in a vacuum tube furnace. The temperature increased from room temperature (T_{room}) to 400 °C within 300 min, and the temperature was held for 15 min to remove the carbamide particles in part-I; then, the temperature continued to increase to 850 °C within 45 min, and the temperature was held for 2 h to sinter the Cu powder in part-II. After that, the furnace cooled to room temperature to obtain the porous Cu samples.

The mass volume method was used to calculate the porosity of the porous Cu samples, and their porosity in each spacer content was considered as the average value of ten samples to ensure repeatability and statistical data. The internal structure and cell wall morphology of the porous Cu samples were observed using a scanning electron microscope (FEI Company, Hillsboro, OR, USA). An electronic universal testing machine (Sunstest, Shenzhen, China) was carried out to measure the compression performance of the porous Cu samples with a 2 mm/min of head movement rate at room temperature. The mechanical properties of porous Cu were given as the average value of five samples.

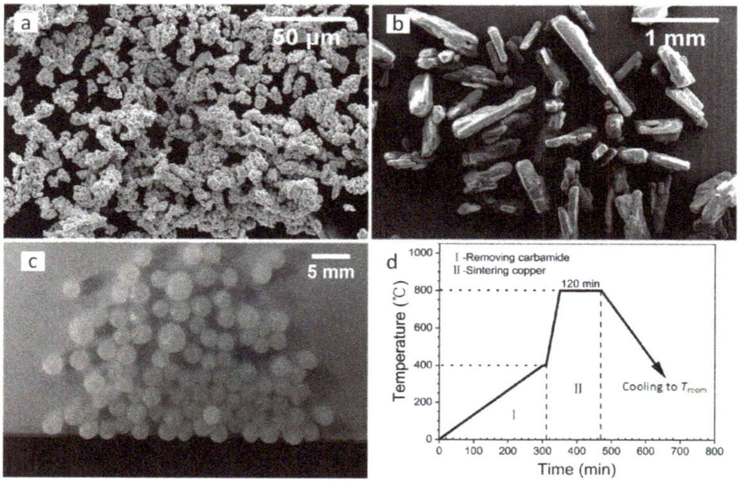

Figure 1. Images of the powder and space holder: (**a**) Cu and (**b**) needlelike and (**c**) spherical carbamide. (**d**) The one step of heat treatment for the fabrication process of the porous Cu.

3. Results and Discussion

3.1. Pore Structure

Figure 2 shows the digital images of the porous Cu samples obtained using different shapes of carbamide particles with a volume content between 10 and 80%, where the upper shows needlelike porous Cu (NP-Cu), and the lower shows spherical porous Cu (SP-Cu). The average porosity of each porous Cu content is shown in Table 1. It can be seen that the porosity of the porous Cu samples increased with the increase in the spacer content. However, the effect of the shape of the space holder on the porosity did not show obvious regularity.

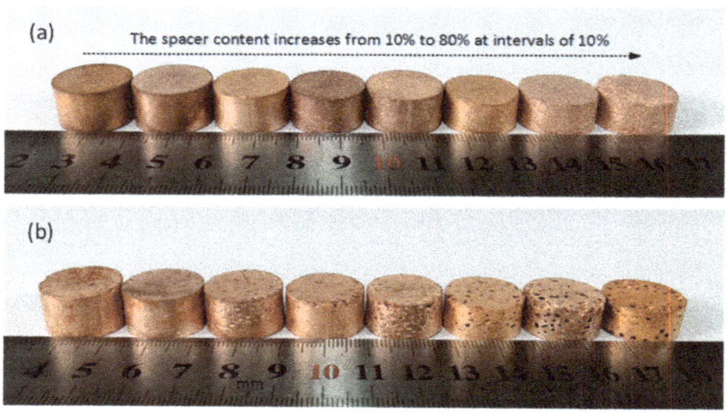

Figure 2. Digital images of the porous Cu samples fabricated using different shapes of carbamide as the space holder with a volume content between 10 and 80%: (**a**) needlelike and (**b**) spherical.

Table 1. Porosity of the obtained NP-Cu and SP-Cu samples in each group.

Spacer Content (%)	10	20	30	40	50	60	70	80
Porosity of NP-Cu/%	26.4 ± 1.5	39.1 ± 0.1	38.7 ± 0.8	42.4 ± 1	57.4 ± 0.5	59 ± 0.6	72.5 ± 0.1	79.3 ± 0.2
Porosity of SP-Cu/%	28.9 ± 1.1	32.1 ± 0.4	39.8 ± 0.9	46.4 ± 0.9	53.8 ± 0.9	62.3 ± 1	71.2 ± 0.1	79.7 ± 0.3

Figure 3 shows the macro- and microtopography of the porous Cu samples in a cross-section. The cutting position diagram of the cross-section is shown in Figure 3a, and the actual cutting surface is shown in Figure 3b, where the upper layer is NP-Cu, and the lower layer is SP-Cu. As can be seen from the figure, the cross-section shows the holes that were not visible on the outer surface. These holes were generated from the removal of the space holder particles and are known as macropores. These macropores are also the main components of porous Cu's porosity. Although the effect on the porosity was not regular, the shape of the space holder had a direct effect on the macropores. That is, the shape of the macropores maintained the shape of the space holder. Since the size of the needlelike space holder of carbamide was 180–380 μm, while the size of the SP-Cu was 0.85–3 mm, the size of the porosity in the case of the porous Cu using a needle-like space holder was smaller than that of SP-Cu.

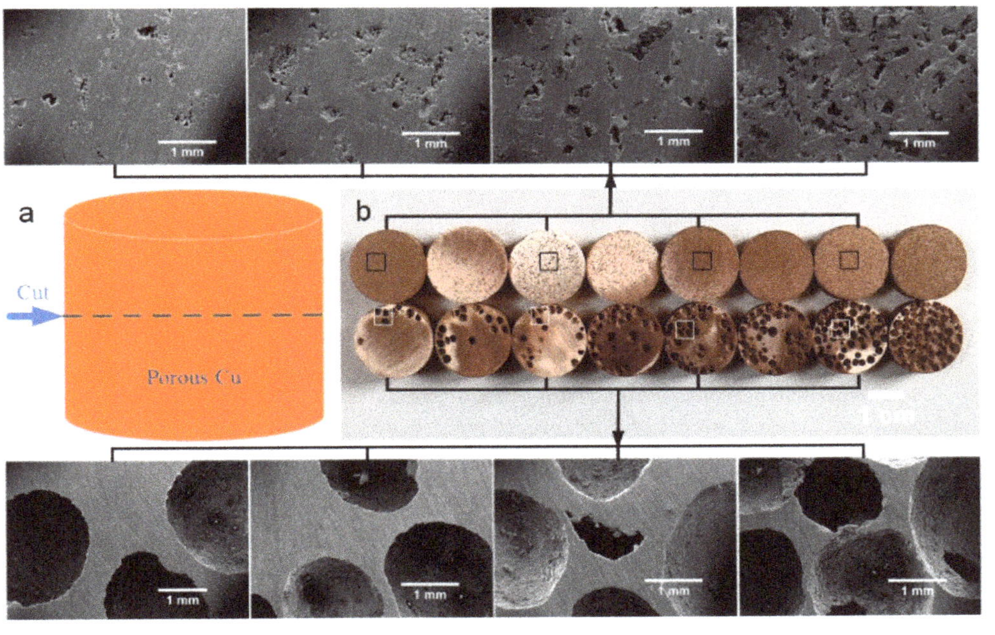

Figure 3. The digital and SEM images of the cross section of the porous Cu samples after being cut parallel to the outer surface at the middle part of the height with molybdenum wire: (**a**) diagram of the cutting site, (**b**) digital image of the cutting surface. The four SEM images above and below are the porous Cu samples prepared with 10%, 30%, 50%, and 70% spacer content.

3.2. Mechanical Properties

Figure 4 shows the compressive stress–strain curves of NP-Cu and SP-Cu with different spacer contents. It can be seen that the stress–strain curve of the porous Cu sample was smooth, showing the deformation characteristics of plastic porous materials. The curve tended to be flat, and the stress level became more obvious with the increase in the spacer content (i.e., porosity). Meanwhile, the bends increased from one at low content to two at high content. Taking the highest content as an example, the curve transitioned from elasticity to plastic deformation at the first turn. At the second bend, the curve transitioned from plasticity to dense deformation. Obvious deformation zones were observed in the porous Cu with high spacer contents. One of the distinct differences between NP-Cu and SP-Cu was that the stress increased differently with the increase in the strain. Compared with the content, the shape of the space holder had less significant influence on the stress–strain curves.

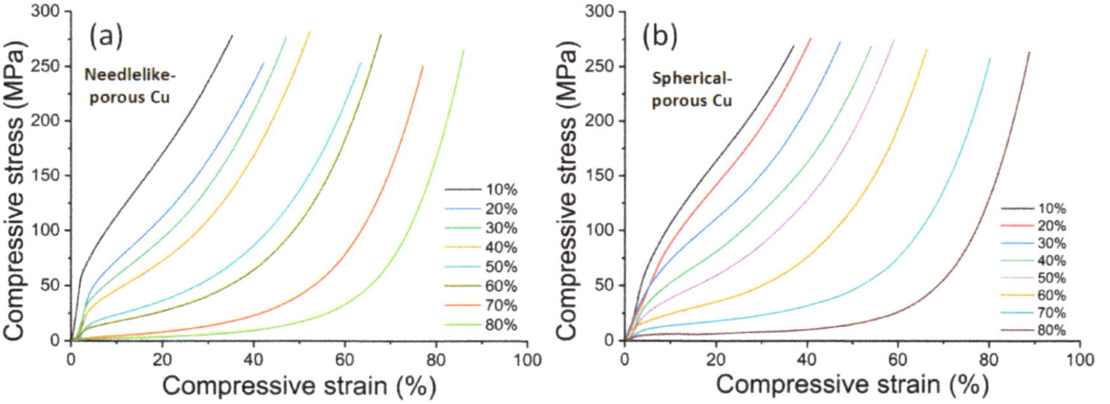

Figure 4. Compressive stress–strain curves of the samples using different shapes of space holders with a volume content between 10 and 80%: (**a**) NP-Cu and (**b**) SP-Cu.

The curves of all the samples showed that the stress increased monotonically with the increase in the strain; that is, there was no obvious yield point and no obvious stress peak. In this way, the elastic modulus was the proportional coefficient in the initial elastic deformation stage, and the yield strength was defined as the stress value of 0.2% residual deformation in the elastic limit [13], as shown in Figure 5, where E and σ_y represent the elastic modulus and yield strength of the porous Cu. These two curves were selected from the above figure for 10% NP-Cu and 80% SP-Cu.

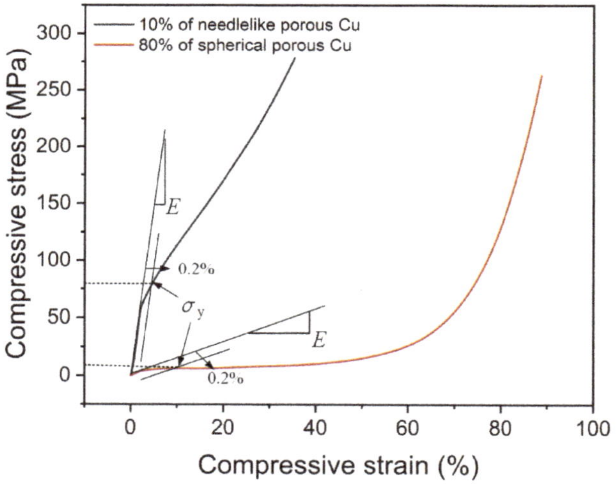

Figure 5. The calculation method of the elastic modulus and yield strength of the porous Cu samples.

The calculated elastic modulus and yield strength of the porous Cu samples are shown in Table 2. It can be seen that the elastic modulus and yield strength tended to decrease with the increase in the spacer content (i.e., porosity). As for the influence of the shape of the space holder, the elastic modulus of the NP-Cu was mostly higher than that of the SP-Cu with the same spacer content instead of 80%, while the yield strength of the SP-Cu was higher than that of the NP-Cu.

Table 2. Elastic modulus and yield strength of the porous Cu samples based on the above method.

Spacer Content (%)	Elastic Modulus (GPa)		Yield Strength (MPa)	
	NP-Cu	SP-Cu	NP-Cu	SP-Cu
10	3.07 ± 0.41	1.58 ± 0.12	63.85 ± 15.02	68.98 ± 4.29
20	2.05 ± 0.36	1.14 ± 0.07	44.78 ± 4.34	71.53 ± 7.99
30	1.58 ± 0.05	1.22 ± 0.08	31.82 ± 0.94	41.85 ± 2.54
40	1.67 ± 0.38	1.05 ± 0.10	25.26 ± 0.77	30.63 ± 2.62
50	1.00 ± 0.20	0.70 ± 0.03	9.21 ± 2.13	23.31 ± 1.49
60	0.59 ± 0.10	0.54 ± 0.03	11.29 ± 0.59	11.66 ± 0.97
70	0.44 ± 0.22	0.43 ± 0.02	3.67 ± 1.30	8.78 ± 0.76
80	0.05 ± 0.01	0.19 ± 0.02	1.85 ± 0.42	4.36 ± 0.32

Figure 6 shows the energy absorption efficiency–compressive stress curves of the porous Cu samples according to the above stress–strain curves. The energy absorption efficiency refers to the ratio of the absorbed energy to the corresponding stress, which was used to determine the best working state of the foam energy absorption [14]. The energy absorption efficiency increased monotonically with the increase in the stress when the spacer content was between 10 and 50%; however, when the spacer content was between 60 and 80%, the energy absorption efficiency first increased and then decreased with the increase in the stress (i.e., peak value). Therefore, the stress corresponding to the peak efficiency (η_{max}) was the densification stress (σ_D), and the strain corresponding to the intensity on the stress–strain curve was the densification strain (ε_D). The densification stress is also called the compressive strength.

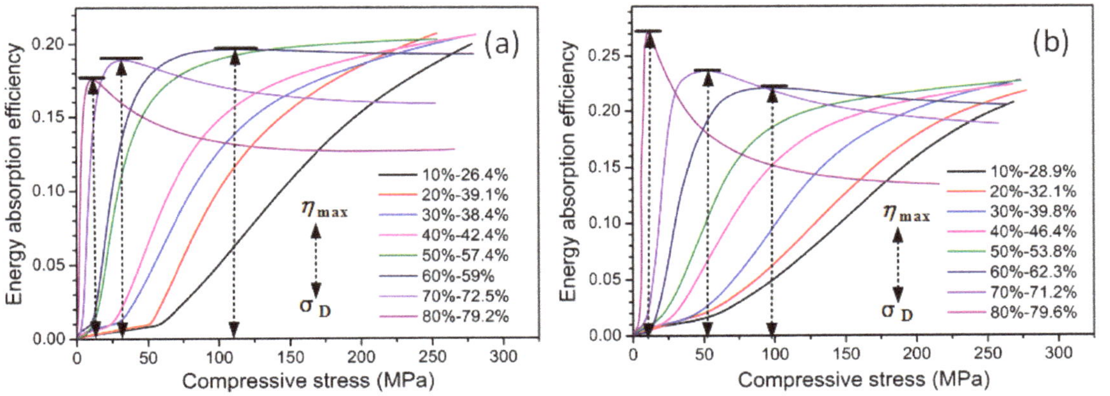

Figure 6. Energy absorption efficiency–compressive stress curves of porous Cu samples fabricated using different shapes of carbamide as the space holder with a volume content between 10 and 80%: (**a**) needlelike and (**b**) spherical. The black arrow indicates the maximum energy absorption efficiency at the upper part and the densification stress at the lower part.

The calculated compressive strength and densification strain of the porous Cu samples with spacer contents between 60 and 80% are shown in Table 3. It can be seen that the compressive strength and densification strain decreased with the increase in the spacer content for the NP-Cu and SP-Cu. This is because the compressive strength decreases with the increase in the porosity, and the porosity increases with the increase in the space holder content; so, the compressive strength decreases with the increase in the space holder content. Due to the decrease in the compressive strength, the porous Cu sample entered the compaction stage earlier, although the stress platform tended to be flat with the increase in the space holder content. Compared with the NP-Cu, the compressive strength and densification strain of the SP-Cu were larger for the spacer contents of 60% and 70%, while they were smaller for the spacer content of 80%. Thus, the effect of the space holder shape on the compressive strength and densification strain of the porous Cu did not show a constant change rule.

Table 3. Compressive strength and densification strain of the porous Cu samples.

Spacer Content (%)	Compressive Strength (MPa)		Densification Strain (%)	
	NP-Cu	SP-Cu	NP-Cu	SP-Cu
60	117.65 ± 6.34	132.02 ± 35.99	51.35 ± 0.72	53.47 ± 3.57
70	33.26 ± 1.19	49.58 ± 8.15	45.78 ± 1.04	47.98 ± 1.74
80	12.07 ± 0.66	11.79 ± 2.45	45.43 ± 0.45	43.47 ± 5.08

In summary, the effect of the space holder shape on the mechanical properties of the porous Cu samples was not only significant but also regular when the spacer content was between 10 and 70%, ignoring 80%, although the effect of the porosity was not regular. In other words, the mechanical properties of the porous materials were not only affected by the porosity but also the pore shape. In fact, the change in the shape of the space holder also caused a change in the particle size. Therefore, different pore shapes and sizes were obtained with the same content, and the pore size also had an effect on the mechanical properties of porous Cu. Thus, to independently examine the effect of the shape of the space holder, one would need to ensure that the particle size was the same; however, this is very difficult to achieve. This may also explain why there are very few studies in the literature dealing with this area.

3.3. Performance Prediction

Although the shape of the space holder had no consistent effect on the porosity and mechanical properties of the porous Cu, we could still examine the effect of the shape of the space holders on the fitting equation through the relationship between them. Generally, the power index relation is widely used to describe the relationship between the elastic modulus and yield strength and the relative density of a porous metal or metal foams [10], which is also called the modified Gibson-Ashby (G-A) model. Figure 7 shows the relationships between the relative elastic modulus and yield strength and the relative density of the porous Cu samples based on the modified and standard G-A models, where E_s and σ_{ys} represent the elastic modulus and yield strength of the solid metal, respectively, and E/E_s and σ_y/σ_{ys} represent the relative elastic modulus and yield strength, respectively. 1-P represents the relative density, C_1 and C_2 represent the coefficients, and n_1 and n_2 represent the exponents. σ_y is often replaced by the plateau stress (the flow stress of a plateau region in the compressive stress-strain curve) of a porous material [15]. The standard G-A model states that the relative elastic modulus of open porous metals is directly proportional to the square of the relative density, and the relative yield strength is directly proportional to the 1.5 square of the relative density [16,17].

It can be seen that the fitting degree of the red lines (i.e., modified G-A model) was significantly higher than that of the black lines (i.e., standard G-A model), which indicates that the modified model was more suitable for predicting the elastic modulus and yield strength of the present porous Cu than the standard G-A model. As far as the authors know, most of the previous studies have shown the same results (see Table 4).

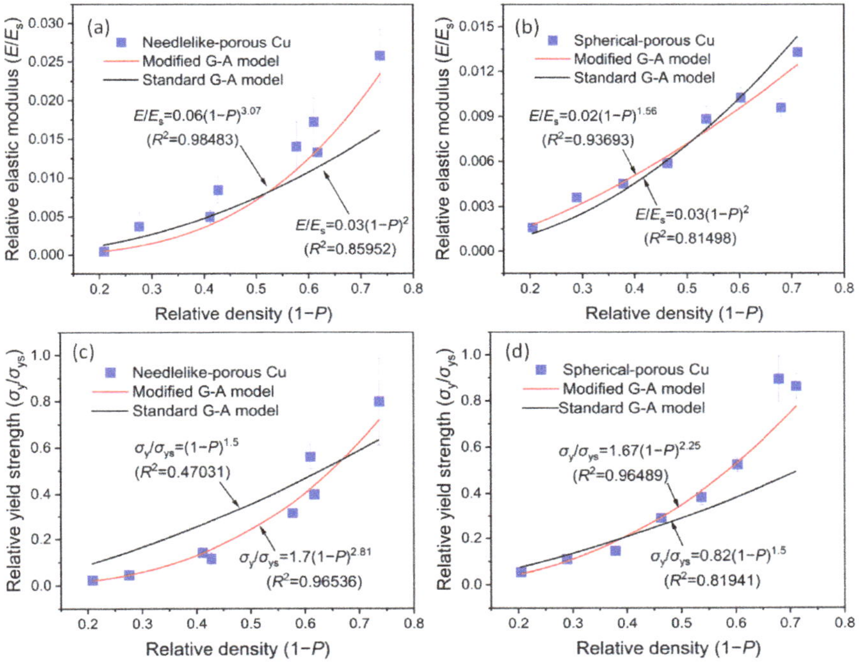

Figure 7. Plots of (**a**,**b**) the relative elastic modulus (E/E_s) and the (**c**,**d**) relative yield strength (σ_y/σ_{ys}) versus the relative density ($1 - P$) with the experimentally obtained data, with standard and modified G-A model formulas for different kinds of porous Cu samples: (**a**,**c**) needlelike and (**b**,**d**) spherical. Here, E_s = 119 GPa, and σ_{ys} = 80 MPa, for the bulk pure Cu.

Table 4. The relations between the mechanical properties and the porosity of porous metals in the literature.

Literature	Skeleton Metal	Elastic Modulus	Yield Strength	Porosity (%)
Ray [5]	Cu	$E/E_s = (1-P)^{(2.6-4.14)}$	-	38.5–70.4
Hong [18]	Cu	-	$\sigma_y/\sigma_{ys} = 0.6(1-P)^{2.28}$	70.2–76.6
Jana [6]	Cu	$E/E_s = 192.61 - 11.09P + 0.27P^2 - 0.003P^3 + 1.39 \times 10^{-5}P^4$		27.5–67.1
Badwe [19]	Au	$E/E_s = 0.86(1-P)^{2.8}$	-	43–70
Bolzoni [20]	Ti Fe	$E/E_s = 0.79(1-P)^{2.49}$ $E/E_s = 0.90(1-P)^{2.54}$	-	44.3–71.1 38.1–75.6
Jenei [21]	Ti	$E/E_s = 1.3(1-P)^{4.6}$	$\sigma_y/\sigma_{ys} = 10.2(1-P)^5$	51–68.4
Tuncer [22]	Ti	-	$\sigma_y/\sigma_{ys} = 1.07(1-P)^{2.28-2.57}$	44–68
Cheneler [23]	Al	$E/E_s = 7.38(1-P)^{2.71}$ $E/E_s = 5.32(1-P)^{2.92}$	$\sigma_y/\sigma_{ys} = 24.6(1-P)^{2.71}$ $\sigma_y/\sigma_{ys} = 21.81(1-P)^{3.09}$	~80–85%

According to Table 1, the relationship between the porosity (P) and spacer content (x) of the porous Cu samples was $P = 0.73x + 0.19$ (for NP-Cu) and $P = 0.74x + 0.18$ (for SP-Cu), respectively. By substituting the former into the formulas in Figure 7a,c and the latter into the formulas in Figure 7b,d, the relationships between the relative elastic modulus and yield strength and the spacer content of the porous Cu can be shown, according to the following formulas. Based on the above formulas, the prediction of the elastic modulus and yield strength of porous Cu, according to the content of the space holder, was realized.

$$\frac{E}{E_s} = 0.04(0.81 - 0.73x)^{3.07} \text{ (for NP} - \text{Cu)} \quad (1)$$

$$\frac{E}{E_s} = 0.01(0.82 - 0.74x)^{1.56} \text{ (for SP} - \text{Cu)} \quad (2)$$

$$\frac{\sigma}{\sigma_s} = 1.24(0.81 - 0.73x)^{2.81} \text{ (for NP} - \text{Cu)} \quad (3)$$

$$\frac{\sigma}{\sigma_s} = 1.24(0.82 - 0.74x)^{2.25} \text{ (for SP} - \text{Cu)} \quad (4)$$

4. Conclusions

The effects of the space holder shape on the pore structure and mechanical properties were studied herein. The porosity of the porous Cu samples increased with the increase in the spacer content; however, there was no regularity with the change in the space holder shape. Meanwhile, the pore shape of the porous Cu samples remained in the shape of space holder, and the pore connectivity increased with the increase in the porosity. Compared with the SP-Cu, it was easier for the NP-Cu to form an open structure due to the smaller size of the space holder used in the NP-Cu. The compressive stress–strain curves of the porous Cu samples were smooth, which showed the mechanical characteristics of porous plastic metals. The mechanical properties of the porous Cu samples decreased with the increase in the spacer content. Although the effect was not as significant as the content, the effect of the shape of the space holder on the mechanical properties of the porous Cu samples, which showed regularity, cannot be ignored; however, the effect of the shape of the space holder on the porosity of the porous Cu was not regular. The relationships between the relative elastic modulus and yield strength and the relative density were well described by a power law with an exponent of approximately 2–3. The results show that the space holder shape also had an influence on the index of fitting equation. Based on a dual model theory, the relationships between the above mechanical properties and spacer content were established. Thus, the source control of the mechanical properties of the porous Cu by the spacer content was realized.

Author Contributions: Investigation, J.X., Y.H., W.M. and Y.Y.; Writing—original draft preparation, J.X.; Writing—review and editing, J.X.; Supervision, J.X. and G.Q. All authors have read and agreed to the published version of the manuscript.

Funding: This research was funded by the National Natural Science Foundation of China (no. 52074052) and by the Natural Science Foundation of Jiangxi Province (no. 20232BAB214041).

Institutional Review Board Statement: Not applicable.

Informed Consent Statement: Not applicable.

Data Availability Statement: The original contributions presented in the study are included in the article, further inquiries can be directed to the corresponding authors.

Conflicts of Interest: The authors declare no conflicts of interest.

References

1. Lu, X.K.; Zhang, Y.B.; Xiao, S.J.; Zhao, Y.; Li, J.S.; Zhou, J.; Ouyang, Q.B. Effect of structural characteristics on the natural convective heat transfer performance of copper foam. *Appl. Therm. Eng.* **2022**, *204*, 118031. [CrossRef]
2. Banhart, J. Manufacture, characterisation and application of cellular metals and metal foams. *Prog. Mater. Sci.* **2001**, *46*, 559–632. [CrossRef]
3. Zhao, Y.Y.; Fung, T.; Zhang, L.P.; Zhang, F.L. Lost carbonate sintering process for manufacturing metal foams. *Scripta Mater.* **2005**, *52*, 295–298. [CrossRef]

4. Shirjang, E.; Akbarpour, M.R. Influence of severe plastic deformation of Cu powder and space holder content on microstructure, thermal and mechanical properties of copper foams fabricated by lost carbonate sintering method. *J. Mater. Res. Technol.* **2023**, *26*, 5437–5449. [CrossRef]
5. Ray, S.; Jana, P.; Kar, S.K.; Roy, S. Influence of monomodal K_2CO_3 and bimodal K_2CO_3+NaCl as space holders on microstructure and mechanical properties of porous copper. *Mater. Sci. Eng. A* **2023**, *862*, 144516. [CrossRef]
6. Jana, P.; Ray, S.; Goldar, D.; Kota, N.; Kar, S.K.; Roy, S. Study of the elastic properties of porous copper fabricated via the lost carbonate sintering process. *Mater. Sci. Eng. A* **2022**, *836*, 142713. [CrossRef]
7. Luo, H.J.; Zhao, J.H.; Du, H.; Yin, W.; Qu, Y. Effect of Mg Powder's Particle Size on Structure and Mechanical Properties of Ti Foam Synthesized by Space Holder Technique. *Materials* **2022**, *15*, 8863. [CrossRef]
8. Jain, H.; Gupta, G.; Mondal, D.P.; Srivastava, A.K.; Pandey, A.; Srivastava, S.K.; Kumar, R. Effect of particle shape on microstructure and compressive response of 316L SS foam by space holder technique. *Mater. Chem. Phys.* **2021**, *271*, 124924. [CrossRef]
9. Mondal, D.P.; Patel, M.; Jain, H.; Jha, A.K.; Das, S.; Dasgupta, R. The effect of the particle shape and strain rate on microstructure and compressive deformation response of pure Ti-foam made using acrowax as space holder. *Mater. Sci. Eng. A* **2015**, *625*, 331–342. [CrossRef]
10. Xiao, J.; Liu, Y.N.; Li, Y.; Qiu, G.B.; Liu, J.M. The application of model equation method in preparation of titanium foams. *J. Mater. Res. Technol.* **2021**, *13*, 121–127.
11. Xiao, J.; Li, X.; Cai, Y.H.; Li, X.T.; Lin, Y.Q.; Xi, J.T. The application of the model equation method in the preparation of porous copper by using needlelike and spherical carbamide as a space holder. *Powder Metall.* **2023**, *66*, 365–376.
12. Xiao, J.; Li, Y.; Liu, J.M.; Zhao, Q.L. Fabrication and Characterization of Porous Copper with Ultrahigh Porosity. *Metals* **2022**, *12*, 1263. [CrossRef]
13. British Standards Institute. *Mechanical Testing of Metals-Ductility Testing-Compression Test for Porous and Cellular Metals*; British Standards Institute: Geneva, Switzerland, 2011.
14. Miltz, J.; Gruenbaum, G. Evaluation of cushioning properties of plastic foams from compressive measurements. *Polym. Eng. Sci.* **1981**, *21*, 1010–1014. [CrossRef]
15. Hakamada, M.; Asao, Y.; Kuromura, T.; Chen, Y.Q.; Kusuda, H.; Mabuchi, M. Density dependence of the compressive properties of porous copper over a wide density range. *Acta Mater.* **2007**, *55*, 2291–2299. [CrossRef]
16. Gibson, L.J.; Ashby, M.F. *Cellular Solids: Structure and Properties*; Cambridge University Press: Cambridge, UK, 1999.
17. Ashby, M.F.; Evans, A.G.; Fleck, N.A.; Gibson, L.J.; Hutchinson, J.W.; Wadley, H.N.G. *Metal Foams: A Design Guide*; Elsevier: Amsterdam, The Netherlands, 2000.
18. Hong, K.; Kadar, C.; Knapek, M.; Drozdenko, D.; Jenei, P.; Kim, M.Y.; Choe, H.; Mathis, K.; Park, H.; Gubicza, J. Comparison of morphology and compressive deformation behavior of copper foams manufactured via freeze-casting and space-holder methods. *J. Mater. Res. Technol.* **2021**, *15*, 6855–6865. [CrossRef]
19. Badwe, N.; Chen, X.Y.; Sieradzki, K. Mechanical properties of nanoporous gold in tension. *Acta Mater.* **2017**, *129*, 251–258. [CrossRef]
20. Bolzoni, L.; Carson, J.K.; Yang, F. Combinatorial structural-analytical models for the prediction of the mechanical behaviour of isotropic porous pure metals. *Acta Mater.* **2021**, *207*, 116664. [CrossRef]
21. Jenei, P.; Kádár, C.; Szabó, A.; Hung, S.M.; Kuo, C.J.; Choe, H.; Gubicza, J. Mechanical behavior of freeze-cast Ti foams with varied porosity. *Mater. Sci. Eng. A* **2022**, *855*, 143911. [CrossRef]
22. Tuncer, N.; Arslan, G.; Maire, E.; Salvo, L. Influence of cell aspect ratio on architecture and compressive strength of titanium foams. *Mater. Sci. Eng. A* **2011**, *528*, 7368–7374. [CrossRef]
23. Cheneler, D.; Kennedy, A.R. A comparison of the manufacture and mechanical performance of porous aluminium and aluminium syntactic foams made by vacuum-assisted casting. *Mater. Sci. Eng. A* **2020**, *789*, 139528. [CrossRef]

materials

MDPI

Article

Enhancing Capillary Pressure of Porous Aluminum Wicks by Controlling Bi-Porous Structure Using Different-Sized NaCl Space Holders

Hongfei Shen, Asuka Suzuki *, Naoki Takata and Makoto Kobashi

Department of Materials Process Engineering, Graduate School of Engineering, Nagoya University, Furo-cho, Chikusa-ku, Nagoya 464-8603, Japan; shen.hongfei.f4@s.mail.nagoya-u.ac.jp (H.S.); takata.naoki@material.nagoya-u.ac.jp (N.T.); kobashi.makoto@material.nagoya-u.ac.jp (M.K.)

* Correspondence: suzuki.asuka@material.nagoya-u.ac.jp

Abstract: Capillary pressure and permeability of porous media are important for heat transfer devices, including loop heat pipes. In general, smaller pore sizes enhance capillary pressure but decrease permeability. Introducing a bi-porous structure is promising for solving this trade-off relation. In this study, the bi-porous aluminum was fabricated by the space holder method using two different-sized NaCl particles (approximately 400 and 40 μm). The capillary pressure and permeability of the bi-porous Al were evaluated and compared with those of mono-porous Al fabricated by the space holder method. Increasing the porosity of the mono-porous Al improved the permeability but reduced the capillary pressure because of better-connected pores and increased effective pore size. The fraction of large and small pores in the bi-porous Al was successfully controlled under a constant porosity of 70%. The capillary pressure of the bi-porous Al with 40% large and 30% small pores was higher than the mono-porous Al with 70% porosity without sacrificing the permeability. However, the bi-porous Al with other fractions of large and small pores did not exhibit properties superior to the mono-porous Al. Thus, accurately controlling the fractions of large and small pores is required to enhance the capillary performance by introducing the bi-porous structure.

Keywords: capillary pressure; permeability; space holder; bi-porous structure; aluminum foam

Citation: Shen, H.; Suzuki, A.; Takata, N.; Kobashi, M. Enhancing Capillary Pressure of Porous Aluminum Wicks by Controlling Bi-Porous Structure Using Different-Sized NaCl Space Holders. *Materials* **2024**, *17*, 4729. https://doi.org/10.3390/ma17194729

Academic Editors: Yuyuan Zhao and Huiping Tang

Received: 10 September 2024
Revised: 21 September 2024
Accepted: 25 September 2024
Published: 26 September 2024

1. Introduction

Open-cell porous metals are a type of metal foam characterized by interconnected pores that allow for fluid flow and exhibit unique properties such as high specific surface area per volume, fluid permeability, and lightweight structure. These characteristics enhance their performance in various applications such as catalysis, filtration, and thermal management [1–3]. Their versatility extends to fields like medical implants, energy storage, and environmental engineering [4,5]. The ability to customize their porosity makes open porous metals adaptable for specific needs, driving innovation and utility across diverse industries.

Open-cell porous metals are promising wick materials in the loop heat pipes [6–8]. In the LHP, capillary pressure and permeability play crucial roles in the efficient transport of heat. LHPs rely on the capillary action of a porous wick structure to drive the working fluid from the evaporator to the condenser without requiring mechanical pumps. Capillary pressure, generated by the surface tension between the liquid and the solid wick, enables the fluid to overcome gravitational and frictional resistance, ensuring continuous heat transfer. Permeability, a property of the wick material, influences the ease with which the fluid can move through the porous structure. High permeability enhances fluid flow, reducing pressure losses and improving overall thermal performance. Therefore, optimizing both capillary pressure and permeability is essential for enhancing the efficiency and reliability of LHPs in thermal management systems, particularly in space and electronics cooling applications [9,10]. Numerous studies have focused on finding porous structures with high capillary pressure and permeability [11,12]. The pore size is an important factor for the capillary pressure and permeability [13,14]. The following formula expresses the capillary pressure (ΔP_{cap}).

$$\Delta P_{\text{cap}} = \frac{2\sigma_{\text{L}} \cos\theta}{r} \quad (1)$$

Here, σ_{L} is the surface tension of the liquid (the force that causes the surface of a liquid to contract), q is the contact angle (the angle where a liquid meets a solid surface, indicating how well the liquid wets or spreads on that surface), and r is the flow channel radius. r correlates with the pore size in the porous wicks, and the capillary pressure can be enhanced by reducing the pore size. However, the reduced pore size resulted in a large flow resistance and low permeability. The permeability (K) is expressed in the following formula.

$$K = \frac{\varepsilon D_e^2}{32(L_e/L)^2} \quad (2)$$

Here, the e is the porosity, L_{e}/L is the tortuosity (a measure of how convoluted or winding a path is within a material, affecting the ease with which fluids can flow through it), and D_{e} is the flow channel diameter. D_{e} also correlates with the pore size in the porous wicks, and the permeability is decreased by reducing the pore size [15]. Thus, the permeability and capillary pressure have a trade-off relation via the pore size. A solution to the trade-off relation between the permeability and capillary pressure is to introduce bi-porous structures with bi-modal pore size distribution. Many experimental and computational studies have demonstrated that the bi-porous structure exhibited superior capillary performance compared to the mono-porous structure [16–18]. Liquid preferentially infiltrates into smaller pores and wets the cell walls of larger pores, improving the capillary pressure.

Many researchers have attempted to prepare bi-porous materials via combining the space holder method, which has a high controllability of porous structure, and another process. Cao [19,20] fabricated bi-porous Ti_3AlC_2 wicks via sintering Ti/Al/TiC powder mixtures with a sodium chloride (NaCl) space holder, which is dissolvable in water. The bi-porous structure consisted of fine gap pores (<2 μm) and large pores formed by space holders (6–12 μm). Zhu [21] produced hierarchical porous Cu containing both micro- (200–450 μm) and nanopores (200–600 nm) through combining a K_2CO_3 space holder and chemical dealloying [22]. Shu [23,24] manufactured porous NiAl intermetallic compounds with a bi-porous structure via combining the combustion synthesis reaction [25] and the NaCl space holder method [26]. Small pores with several micrometer sizes were formed by the Ni-Al combustion synthesis reaction. However, these processes stochastically form small pores and cannot intentionally control their fraction and sizes, resulting in a poor understanding of suitable bi-porous structures (e.g., fractions of large and small pores). One promising approach for intentionally controlling the bi-porous structure is the space holder method using space holder particles of different sizes. It is expected that the fraction and size of large and small pores can be controlled independently through tailoring the fraction and size of two space holder particles. Evaluating the capillary performance of bi-porous materials controlled by the two different-sized space holder methods leads to a better understanding of suitable bi-porous structures for the porous wicks.

In our previous study [27], sodium chloride (NaCl) powders of around 400 μm were applied as the space holder to prepare porous aluminum (Al) with a different porosity. Increasing porosity led to higher fluid permeability and lower capillary pressure even though the pore size barely changed. This was because connective parts between adjacent pores were widened by increasing porosity. The connective parts had a small flow channel size, and its widening increased the permeability and decreased the capillary pressure. Thus, the porous Al fabricated via the space holder method also exhibited the trade-off relation between the capillary pressure and fluid permeability, which was indirectly affected by the pore fractions. In this present study, bi-porous Al was fabricated using

NaCl space holder particles of two different sizes. The effect of the volume fractions of small and large NaCl particles on the porous structure and the capillary performance was systematically investigated under a constant total volume fraction of NaCl, e.g., constant total porosity. These results were used to discuss suitable bi-porous structures for enhancing capillary performance.

2. Materials and Methods

2.1. Space Holder Method

Figure 1 shows scanning electron microscope (SEM) images of the raw materials and the fabrication process of bi-porous Al using the space holder method. Powders of aluminum ($\leq$45 μm, 99.99%) and NaCl particles with different sizes (30–50 μm and 330–430 μm) were used. Different combinations of Al powder and NaCl particles considered in this study are summarized in Table 1. In the mono-porous group, only small NaCl particles were blended with aluminum so that the NaCl volume fractions were set at 50%, 60%, and 70%. Volume fractions greater than 80% cannot maintain the skeleton after dissolving in water, while lower volume fractions cannot ensure the complete dissolution of NaCl. In the bi-porous group, the total volume fraction of NaCl was kept constant at 70% for the same reason, while the volume fraction of small NaCl particles was increased from 10% to 40%. This also allows us to compare them with the counter mono-porous sample under the same total porosity. The samples were labeled using L (large), S (small), and the volume fractions. For example, the porous sample fabricated using the large NaCl of 60% and the small NaCl of 10% was denoted as L60S10. Al and NaCl were manually blended in a bowl with a ceramic bar for 10 min, twice and placed into a cylindrical graphite mold (inside diameter: 10 mm, outside diameter: 20 mm, height: 50 mm), followed by spark plasma sintering (CSP-KIT-02121, PLASMAN, Hiroshima, Japan) under vacuum conditions (~10 Pa). The temperature was increased at a heating rate of approximately 0.5–0.67 °C/s and kept at 600 °C for 600 s under a constant applied pressure of 20 MPa. The sintered samples were placed in running water for over 86.4 ks (24 h) to completely remove the NaCl particles. The samples were put into a hot dryer and heated at 150 °C for 10.8 ks (3 h).

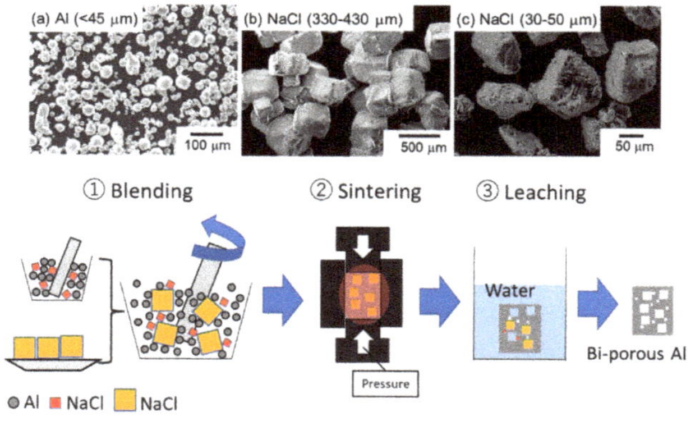

Figure 1. SEM images of raw materials and fabrication process of bi-porous Al using space holder method.

The mass, diameter, and height of each sample were measured to quantify the total porosity (ε) [28].

$$\varepsilon = 1 - \frac{\rho_{\text{sample}}}{\rho_{\text{Al}}} = 1 - \frac{4m_{\text{leached}}}{\pi d^2 h \rho_{\text{Al}}} \tag{3}$$

Here, ρ_{sample} is the bulk density of the sample after NaCl removal, m_{leached} is the mass of the samples after NaCl removal, ρ_{Al} is the true density of Al (2.7 g cm^{-3}), and d and h are the diameter and height of the cylindrical sample (d = 10 mm, h = 19~23 mm). The quantified porosity is summarized in Table 1 and is well consistent with the designed total porosity of each sample. The porous samples were immersed in epoxy resin and polished with SiC abrasive papers and diamond slurries of 3 μm and 1 μm in particle sizes. The porous structures were observed using an SEM (JSM-IT500, JEOL, Tokyo, Japan). Image analysis using software (Image J 1.53t, National Institutes of Health, Bethesda, MD, USA) was conducted to estimate porosity and pore size.

Table 1. The blending volume fraction of large and small NaCl particles for fabricating mono-porous and bi-porous samples. The total porosity quantified by measuring the mass, diameter, and height of each sample is also shown.

	Sample	330~430 μm NaCl Volume Fraction (%)	30~50 μm NaCl Volume Fraction (%)	Designed Total Porosity (%)	Measured Total Porosity (%)
Mono-porous	S50	0	50	50	48.9
	S60	0	60	60	59.9
	S70	0	70	70	68.7
Bi-porous	L60S10	60	10	70	70.5
	L50S20	50	20	70	72.3
	L40S30	40	30	70	72.6
	L30S40	30	40	70	72.0

2.2. Capillary Rise Experiment

Figure 2 shows a schematic of the experiment setup. The cylindrical sample (diameter: 10 mm, height: 19–23 mm) was contacted vertically to the water reservoir, and then the water was spontaneously infiltrated upward into the porous sample. Infrared (IR) thermal imaging was utilized to capture the water height during the capillary rise processes. The infrared camera (TVS-500EX, Nippon Avionics, Yokohama, Japan) with a temperature resolution better than 0.05 K and an accuracy of 2% for temperatures below 300 °C was used. The detailed measurement method is given elsewhere [27]. Before the experiments, all the test samples were dried for 1.8 ks (0.5 h). The tests were conducted at an ambient temperature of 22 ± 1 °C. To reduce test errors, each sample was tested three times, and the average value was taken as the final result.

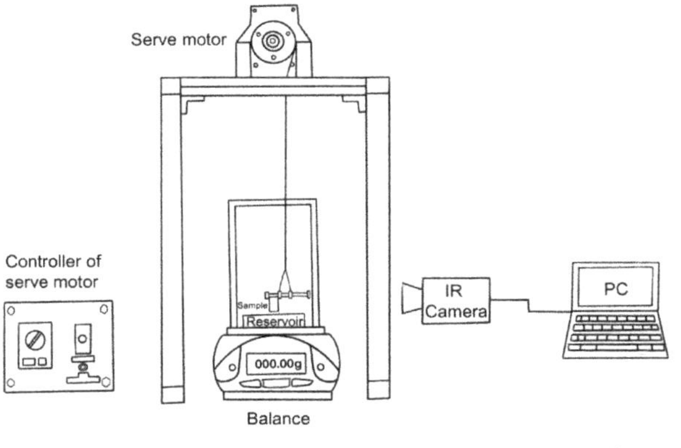

Figure 2. Schematic illustration of setup for measuring the capillary performance of porous samples in this study [27].

According to Darcy's law, liquid is driven by capillary pressure and countered by viscous resistance and gravity in the capillary wicking process [29].

$$\Delta P_{\text{cap}} = \frac{\mu\varepsilon}{K} h \frac{dh}{dt} + \rho g h \tag{4}$$

In Equation (4), the first term on the right-hand side refers to viscous resistance based on Darcy's law while the second term represents the effect of gravity. g is the gravitational acceleration. Equation (4) can be modified as follows:

$$\frac{dh}{dt} = \frac{\Delta P_{cap} K}{\mu\varepsilon} \cdot \frac{1}{h} - \frac{\rho g K}{\mu\varepsilon} \tag{5}$$

Equation (5) shows that there is a linear correlation between the capillary rising rate ($\mathrm{d}h/\mathrm{d}t$) and the reciprocal height of liquid (h^{-1}). Thus, when the measured correlation between $\mathrm{d}h/\mathrm{d}t$ and h^{-1} is regressed linearly, K and $\Delta P_{\text{cap}} \cdot K$ can be quantified from the intercept and slope of the regression line. ΔP_{cap} can be evaluated by dividing $\Delta P_{\text{cap}} \cdot K$ by K.

3. Results

3.1. Mono-Porous Al with Various Porosity

Figure 3 displays the SEM images of the microstructures of mono-porous Al with various volume fractions of small NaCl. The dark areas indicate the resin infiltrating into the pores and the bright areas represent the Al part. The shape of the pores was different from the original shape of the small NaCl particles (30–50 μm). The Al cell wall exhibited a particulate morphology, indicating that sintering was inhibited. Comparing with our previous work [27], where the large NaCl (330–430 μm) were used, the shape of the pores replicated the shape of the large NaCl particles, and the dense Al cell wall was formed. It was suggested that the particle size ratio of NaCl and Al (<45 μm) powders affected the pore morphology. When the NaCl particles were much larger than Al particles, the Al particles surrounded the NaCl particles and easily contacted adjacent Al particles, resulting in the forming of a dense cell wall via spark plasma sintering. In contrast, when the NaCl particle size was comparable to Al particles, the contacts between the Al particles were inhibited by larger fractions (50–70%) of NaCl particles. As a result, the Al cell wall was particulate, and the pores did not replicate the shape of NaCl particles. Even in sample S50, fabricated using small NaCl particles at a 50% volume fraction, it was confirmed through mass measurement that space holders were completely removed. As the volume fraction of NaCl increased from 50% to 70%, more small pores interconnected with each other. Consequently, the average pore size measured via image analysis increased from 35.2 to 148.3 μm. These results indicated that when the NaCl particle size was similar to the Al powder size, changing the volume fraction of NaCl could not only control the porosity but also vary the pore size.

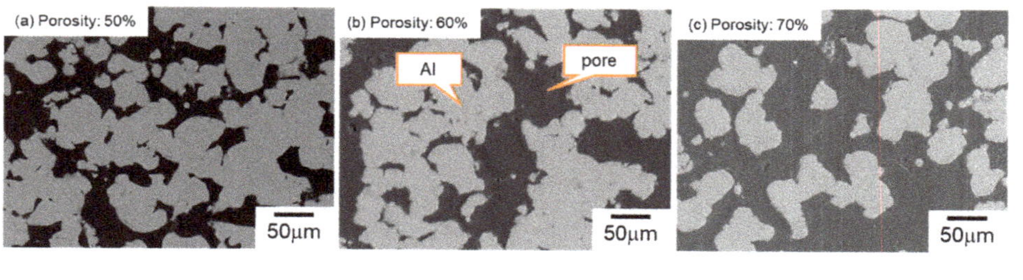

Figure 3. Representative SEM images of mono-porous Al: (**a**) S50, (**b**) S60, and (**c**) S70.

Figure 4a presents the time evolution of the capillary rise height captured by the IR camera. In all the samples, the infiltrated water height sharply increased at the early stage of infiltration because of a small gravity effect. The rising rate slowed down as the influence of gravity became significant. The sample with larger porosity had a faster capillary rising rate. Infiltrated water reached the sample height within 15 s in sample S70. In samples S60 and S50, the water height continuously increased for 30 s and did not reach the sample height or equilibrium infiltration height at which the capillary force was balanced with the gravity during the measurement period.

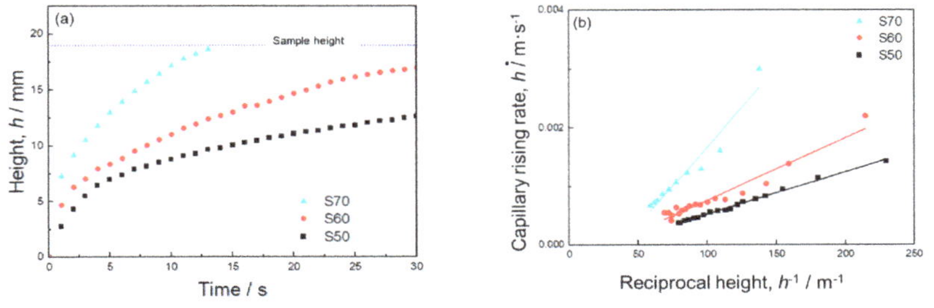

Figure 4. (**a**) Time evolution of the capillary rising height of mono-porous Al. (**b**) Relationship between capillary rising rate and reciprocal height.

Figure 4b shows the relationship between the capillary rise rate (dh/dt) and the reciprocal height (h^{-1}). In this figure, the data point at the final stage of the infiltration into sample S70 were excluded to eliminate the effect of water reaching the sample height on the capillary rise rate. There are good linear correlations between dh/dt and h^{-1}, indicating that the capillary pressure and permeability can be quantified using Equation (5).

The calculated values of ΔP_{cap}, K, and $\Delta P_{cap} \cdot K$ are provided in Table 2. K was much more sensitive to porosity than ΔP_{cap}: with increasing the porosity from 50% to 70%, K increased by approximately 7 times (1.1×10^{-11} m^2 to 7×10^{-11} m^2) while ΔP_{cap} decreased by 12.5% (360 to 315 Pa). Consequently, the capillary performance ($\Delta P_{cap} \cdot K$) was enhanced by approximately 5 times when the porosity increased from 50% to 70%. As mentioned above, increasing porosity from 50% to 70% increased the pore size from 35.2 μm to 148.2 μm (Figure 3). According to Equation (1), the larger pore size reduced the capillary pressure. However, based on Equation (2), the permeability was improved by increasing the porosity and pore size. Thus, mono-porous Al fabricated using small NaCl particles exhibited the trade-off relation between the capillary pressure and permeability via the porosity. This trend was consistent with the results of mono-porous Al fabricated using large NaCl particles [27].

Table 2. Capillary performance of porous Al obtained in this study.

Sample	Average Large Pore Size (μm)	Average Small Pore Size (μm)	Permeability, $K/10^{-11}$ m^2	Capillary Pressure, ΔP_{cap}/Pa	Capillary Factor, $\Delta P_{cap} \cdot K/10^{-9}$ N
S50	-	35.2	1.1 ± 0.1	360 ± 38.0	4.0
S60	-	53.5	2.5 ± 0.6	336 ± 13.9	8.4
S70	-	148.3	7 ± 1.6	315 ± 31.6	22
L70	270	-	10.4 ± 3.1	197 ± 15.4	20.5
L60S10	228	22.1	9.3 ± 1.2	265 ± 19.6	24.6
L50S20	260	51.1	8.7 ± 1.3	281 ± 17.8	24.4
L40S30	276	65.7	8.4 ± 1.7	381 ± 28.0	32
L30S40	274	96.6	7.4 ± 1.1	293 ± 12.7	21.7

3.2. Bi-Porous Al with Various Volume Fractions of Large and Small Pores under a Constant Total Porosity

Figure 5 shows the SEM images of the microstructures of bi-porous Al denoted as samples (a) L6010, (b) L50S20, (c) L40S30, and (d) L30S40. A typical bi-porous structure was successfully fabricated using NaCl space holder particles of two different sizes. The total porosity measured (listed in Table 1) was around 70%, which confirmed the removal of all the NaCl particles. In low-magnification images, cuboidal pores replicating the shape of large NaCl particles (Figure 1) were observed. The cuboidal large pores were less observed and became isolated when the volume fraction of large NaCl particles was decreased. In high-magnification images, although the shape of the small pores was not cuboidal, the size of the small pores was comparable to that of the small NaCl particles. When the volume fraction of small NaCl particles was increased, the small pores interconnected with each other. The area fraction of small pores in the cell wall of large pores was measured using image analysis and quantified at 23.4% for L60S10, 42.7% for L50S20, 47.6% for L40S30, and 52.8% for L30S40. The designed volume fraction of small pores in the cell wall was 25%, 40%, 50%, and 57%, respectively. The measured area fractions of the small pores were almost consistent with the designed volume fraction. These results demonstrated a high controllability of bi-porous structure using NaCl space holder particles with different particle sizes.

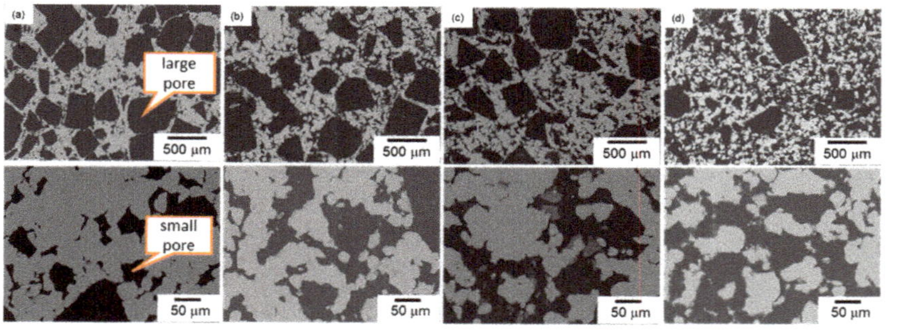

Figure 5. SEM images of bi-porous Al: (**a**) L60S10, (**b**) L50S20, (**c**) L40S30, and (**d**) L30S40.

Figure S1 shows SEM images of feature pore structures of bi-porous Al. The small pores in the cell wall acting as the interconnected channels between the big pores can be observed, which guarantees an open-cellular structure between the big pores and helps improve the capillary performance of the bi-porous structure significantly.

Figure 6a shows the time evolution of the capillary height of four bi-porous samples measured by the IR camera. For comparison, two mono-porous samples, L70 [27] and S70 (Section 3.1), with the same total porosity as the bi-porous samples, are also presented in this figure. All the samples exhibited typical infiltration height curves, on which the infiltration height increased sharply at first and then slowly with infiltration progress. Finally, the infiltration height tended to saturate as the water reached the top of the sample. For the four bi-porous samples with constant total porosity, the increase in the fraction of small pores did not monotonously enhance the rising rate. The sample L60S10 exhibited a much higher capillary rising rate than the sample L70, indicating that replacing 10% large pores with small pores enhanced the capillary performance. The capillary rising rate of sample L50S20 was almost comparable to that of sample L60S10. Sample L40S30 exhibited a further improved capillary rising rate compared to samples L50S20 and L60S10. However, the capillary rising rate of sample L30S40 was smaller than that of sample L40S30 and was almost comparable to sample S70. These results indicated that bi-porous structures did not always exhibit better capillary performance than mono-porous structures and needed to be optimized.

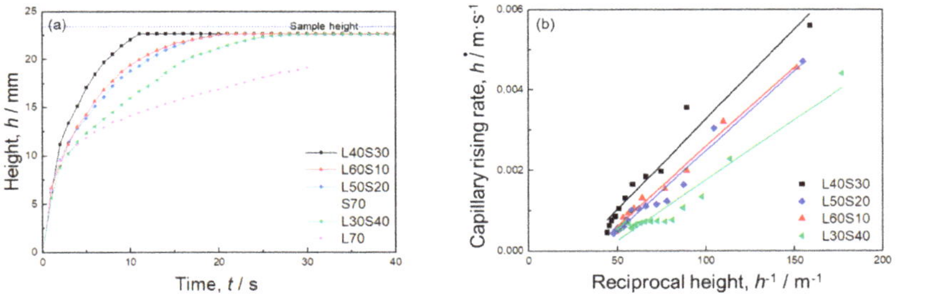

Figure 6. (**a**) Time evolution of the capillary rising height of bi-porous Al. (**b**) Relationship between capillary rising rate and reciprocal height.

Figure 6b shows the relationship between the capillary rising rate (dh/dt) and reciprocal height (h^{-1}) of the bi-porous samples. In the cases of samples L40S30 and L60S10, there are almost linear correlations between dh/dt and h^{-1} (R^2 = 0.96 for L40S30 and 0.99 for L60S10). The capillary pressure and permeability of these samples can be obtained based on Equation (5). Samples L50S20 and L30S40 exhibited non-linear correlations between dh/dt and h^{-1}, suggesting that the capillary performance changes with the infiltration progress. However, the deviation from regressed lines was small (R^2 = 0.97 for L50S20 and 0.93 for L30S40). Analyzing the regression lines based on Equation (5) can provide an average of the capillary pressure and permeability throughout the overall infiltration into these samples. The quantified values of ΔP_{cap}, K, and $\Delta P_{cap}{\cdot}K$ are provided in Table 2.

Figure 7a shows changes in the capillary pressure and permeability with the volume fraction of small NaCl particles (corresponding to the volume fraction of small pores). The capillary pressure first increased when the volume fraction of small pores increased from 0% to 30% and decreased when the volume fraction of small pores was 40%. Compared to sample L30S40, sample S70 exhibited a slightly higher capillary pressure. As a result, the sample L40S30 exhibited the highest capillary pressure of 381 Pa. The permeability kept decreasing slightly from 10.4×10^{-11} m^2 to 7×10^{-11} m^2 as the volume fraction of small pores increased. Thus, tailoring the bi-porous structure using two-sized space holder particles enhanced the capillary pressure while minimizing the decrease in permeability.

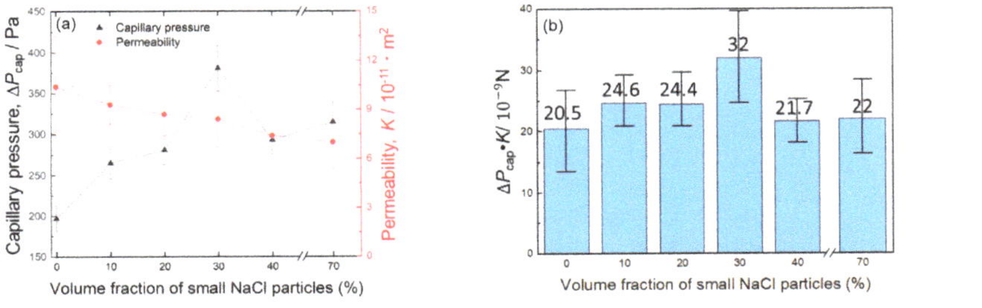

Figure 7. Change in (**a**) permeability, capillary pressure, and (**b**) their product with the volume fraction of small NaCl particles.

The capillary performance ($\Delta P_{cap}{\cdot}K$) was also evaluated, as shown in Figure 7b. Samples L40S30 presented the highest $\Delta P_{cap}{\cdot}K$, followed by L50S20, L60S10, S70, L30S40, and L70. This trend was consistent with the trend in capillary height curves in Figure 6a. Samples L40S30, L50S20, and L60S10 exhibited higher $\Delta P_{cap}{\cdot}K$ than mono-porous samples L70 and S70. It was confirmed that the bi-porous structure can improve capillary performance. In particular, the optimized sample L40S30 exhibited approximately 1.5 times higher capillary performance than the mono-porous samples.

Figure S2 displays the size distribution of small pores on cell walls of the large pores with different volume fractions of small NaCl. When the volume fraction was 10% (Figure S2a), most of the pores had smaller than 20 μm, but pores of sizes > 50 μm also existed. On the other hand, when the volume faction was 20% (Figure S2b) and 30% (Figure S2c), pore size had a concentrated distribution in 20–60 μm, and pores of sizes > 100 μm were formed. Thus, as the volume fraction of small NaCl increased, the small pore size increased and the distribution apparently widened and shifted to the large size. However, the effect of size distribution on the capillary performance cannot be discussed from these results because the volume fraction of large and small pores and the average pore size were changed at the same time. To discuss the effect of the pore size distribution on the capillary performance, it is necessary to fabricate the samples with various pore size distribution and a constant porosity and average pore size. This can be conducted in future works.

In Figure 8, the capillary pressure and permeability of all the samples evaluated in this study are plotted. For comparison, the properties of mono-porous Al fabricated using the large NaCl particles [27] are also shown in this figure. The dotted lines indicate isolines of the capillary performance ($\Delta P_{cap} \cdot K$). As mentioned in Section 3.1, mono-porous samples exhibited the trade-off relationship between ΔP_{cap} and K via the porosity. When the porosity was the same, decreasing space holder size (pore size) increased the capillary pressure and decreased the permeability, resulting in almost constant capillary performance regardless of the pore size (compare L70 with S70 and L60 with S60). This trend was a well-known trade-off relation between ΔP_{cap} and K via the pore size.

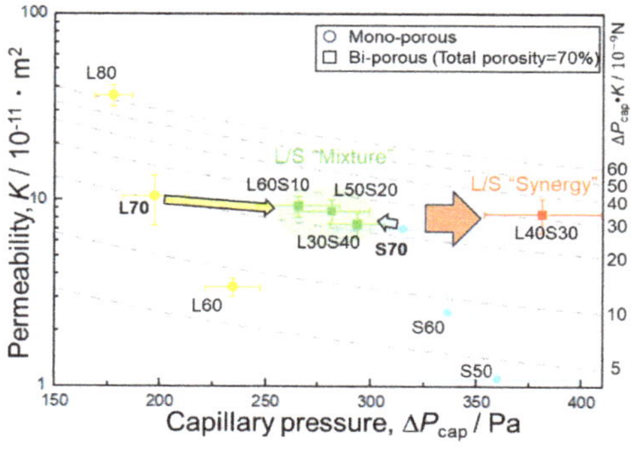

Figure 8. Plot of permeability and capillary pressure of mono-porous and bi-porous Al.

When the bi-porous samples were focused on, the capillary pressure and permeability of samples L60S10, L50S20, and L30S40 ranged from 265 to 293 Pa and 7.4×10^{-11} m^2 to 9.3×10^{-11} m^2, respectively, which were located between those of samples L70 ($\Delta P_{cap} = 197$ Pa, $K = 10.4 \times 10^{-11}$ m^2) and S70 ($\Delta P_{cap} = 315$ Pa, $K = 7 \times 10^{-11}$ m^2). Roughly speaking, these samples can be considered as mixtures of samples L70 and S70, although $\Delta P_{cap} \cdot K$ of the samples L60S10 and L50S20 was slightly higher than that of mono-porous samples. In contrast, sample L40S30 deviated far from the mixture-like properties. The capillary pressure of sample L40S30 ($\Delta P_{cap} = 381$ Pa) was much higher than those of samples L70 and S70 while maintaining the permeability ($K = 8.4 \times 10^{-11}$ m^2). It was considered that the large and small pores exhibited a synergy to enhance the capillary pressure. Thus, the bi-porous structure can overcome the trade-off relationship between ΔP_{cap} and K but must be tailored so that the properties do not fall into mere mixture law.

4. Discussion

This present study fabricated the bi-porous Al using two different-sized NaCl space holder particles and evaluated its capillary performance. The effect of the volume fraction of small/large NaCl particles (corresponding to the porosity of small/large pores) on capillary pressure and permeability was investigated under a constant total porosity of 70%. The properties of bi-porous Al were compared to those of mono-porous Al fabricated using the small or large NaCl particles alone. Mono-porous Al exhibited the trade-off relation between the capillary pressure and permeability via the porosity (Table 2 and Figure 8). This was because increasing the NaCl volume fraction improved the connectivity between adjacent pores and formed a large network of pores (Figure 3). The large network worked as large pores (Table 2), resulting in the enhancement of the permeability but decreasing the capillary pressure. The bi-porous structure could be controlled by tailoring the volume fractions of small/large NaCl particles (Figure 5, Tables 1 and 2). Tailoring the porosity of small/large pores under the constant total porosity had a significant effect on the capillary pressure and a slight effect on the permeability (Figure 7a). Although samples L60S10, L50S20, and L40S30 exhibited higher capillary performances than samples L70 and S70 (Figure 7b), samples L60S10 and L50S20 roughly fell into the mixture of samples L70 and S70. Sample L40S30 exhibited the highest capillary performance (Figure 7b) in this study and the synergy effect of samples L70 and S70 on the capillary pressure (Figure 8). The reason for the better properties of sample L40S30 is discussed here.

In Figure 9, average large, small, and overall pore sizes are plotted as a function of the volume fraction of small NaCl particles under a constant total volume fraction of NaCl of 70%. In addition, the effective flow channel size was inversely quantified using the measured capillary pressure (Table 2) and Equation (1). The surface tension of water (s_L) at room temperature was 72×10^{-3} N/m, and the contact angle (q) between the water and Al was reported in the literature as 81° [30]. As the porosity of the small pores increased, the average large pore size remained nearly constant at around 260 μm while the average small pore size increased monotonously from 22.1 to 148.3 μm. As a result, the average overall pore size decreased by introducing small pores, became almost constant in the bi-porous samples, and decreased in sample S70. The effective flow channel size was also located between the average large and small pore sizes and varied with the same trend as the overall average pore size. This trend suggested that capillary pressure changed depending on the overall average pore size, which was consistent with L70/S70 mixture-like properties shown in Figure 8. However, the effective flow channel size of sample L40S30, which exhibited the highest capillary pressure (Figure 8), deviated from this trend. This suggested that the overall average pore size could not determine the capillary pressure of this sample, which corresponded to the noteworthy property compared to the other bi-porous sample shown in Figure 8.

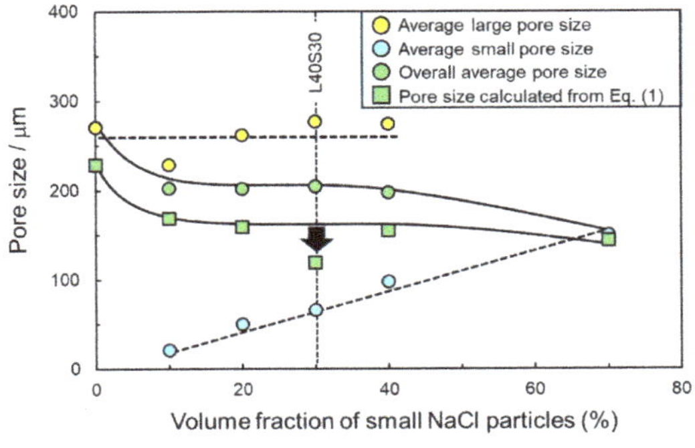

Figure 9. Changes in the average sizes of large, small, and overall pores as a function of the volume fraction of small NaCl particles. The flow channel size calculated from Equation (1) is also shown in this figure.

Byon and Kim reported that the bi-porous wicks, which were fabricated by the sintering process using clusters of small-sized glass particles, exhibited better capillary performance than the mono-porous wicks [18]. They discussed the importance of the ratio of cluster size to particle size. When the ratio is in the range of 4–6, the capillary performance is improved because a liquid is wicked into small pores at first and improves the wettability between the liquid and solid (containing liquid in the small pores) at the large pores. When the ratio is lower or higher than this range, the capillary behaviors are close to the mono-porous wicks with large or small pores. The small pores in sample L40S30 might improve the wettability between water and solid cell walls of large pores and the capillary pressure. In Figure 9, the effective flow channel size was estimated from Equation (1), assuming that the contact angle between water and Al was constant at 81°. However, if wettability at the large pores was improved by water infiltration into small pores, the contact angle was not necessarily constant. The improved contact angle led to underestimating the effective flow channel size as shown in Equation (1).

The ratios of large pore size to small pore size were 10.3 for L60S10, 5.1 for L50S20, 4.2 for L40S30, and 2.8 for L30S40 (Table 2). Assuming that the ratio of cluster size to particle size can be replaced with the ratio of large pore size to small pore size, the size ratio of L50S20 and L40S30 was in the range of the appropriate ratio (4–6), supporting the better property of L40S30. Another important aspect of the bi-porous structure is the connectivity of two different-sized pores. In this study, the NaCl space holders were completely removed in all the samples, indicating the well-connected porous structure. However, the small pores need to be connected with both adjacent small and large pores to wet the cell wall of large pores and improve the wettability. The large pores also need to be connected to each other to ensure the main infiltration path has high permeability. These situations were not always established in all the samples. Specifically, the volume fraction of small pores in the cell wall of large pores in sample L60S10 was approximately 23.4%. The connection of the small pores with adjacent small and large pores was difficult to establish under such a small porosity. The volume fraction of large pores was approximately 30% in sample L30S40, in which the connection between adjacent large pores is difficult to establish. In addition, the capillary rising rate in small pores needs to be higher than that in large pores to improve the wettability because water infiltrating into the smaller pores must precede and wet the solid cell walls of the larger pores. The capillary rising rate at the early stage of infiltration is dominated by $\Delta P_{cap} \cdot K$ in Equation (1). As shown in Figure 8, the $\Delta P_{cap} \cdot K$ of the mono-porous samples with the same porosity (L70/S70 and L60/S60) was almost the same. Assuming that this situation was established in the bi-porous samples, the volume fraction of small pores in the cell wall needs to be higher than that of large pores. The volume fractions of small pores in the cell wall were 23.4% for L60S10, 42.7% for L50S20, 47.6% for L40S30, and 52.8% for L30S40. In samples L40S30 and L30S40, the volume fractions of small pores in the cell wall were higher than that of large pores, suggesting the preceding water infiltration into small pores. Thus, the superior properties of only the sample L40S30 were attributed to the appropriate pore size ratio, connectivity of small pores with both large and small pores, and the volume fraction of small pores in the cell wall higher than that of large pores. Figure 10 shows hypothetical schematic illustrations of capillary rising behaviors of bi-porous samples (a, d) L60S10, L50S20, (b, e) L40S30, and (c, f) L30S40. In samples L60S10 and L50S20, the large pores were well-connected to each other, whereas most of the small pores were connected with either adjacent large or small pores. Then, preceding water infiltration into the large pores was dominant for the capillary rising, and the samples did not benefit from the wetting of the cell walls of the larger pores via water infiltrating into smaller pores (Figure 10a). Based on this hypothesis, the situation is simplified, such as in Figure 10d, in which water infiltrates into large and small pores independently. In sample L30S40, the small pores became well-connected to each other, whereas the large pore network was separated. Then, water preferentially infiltrated into the small pores (Figure 10c). Although water infiltrating into the small pores improved the wettability at the large pores, poor connectivity between adjacent large pores resulted in infiltration into the large pores being significantly delayed due to infiltration into the small pores. Following this assumption, the situation is simply illustrated in Figure 10f. In sample L40S30, not only were the large pores well-connected to each other, but the small pores were also connected with adjacent large and small pores, allowing infiltration into both pores (Figure 10b). The slightly higher volume fraction of small pores in the cell wall (47.6%) compared to the volume fraction of large pores (40%) produced moderately preferential water infiltration in the small pores. Water infiltrating into the smaller pores then promoted improved wettability and meniscus formation within the larger pores, improving the capillary pressures within the larger pores. This situation is simplified in Figure 10e. However, these mechanisms were not completely demonstrated. To elucidate the mechanism of the superior properties in more depth, it is required to evaluate the capillary performance of the samples in which the bi-porous structures are further controlled (e.g., various fractions of large and small pores under a constant small pore size) and to observe the capillary rising behaviors directly.

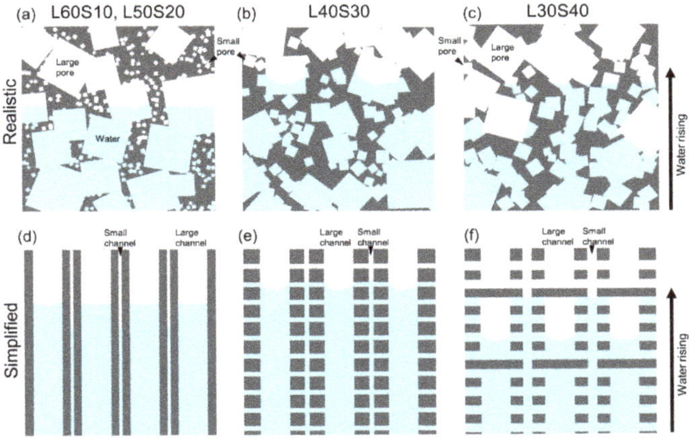

Figure 10. Hypothetical schematic illustration of capillary rising behaviors in (**a**) L60S10, L50S20, (**b**) L40S30, and (**c**) L30S40. The complex porous structures are simplified into straight channel models in (**d**–**f**).

5. Conclusions

In this study, bi-porous structures were controlled by sintering aluminum (Al) powder with sodium chloride (NaCl) space holder particles of two different sizes. The capillary pressure (ΔP_{cap}), permeability (K), and their product ($\Delta P_{cap} \cdot K$) were investigated and compared to that of mono-porous Al. The major conclusions are as follows:

(1) Mono-porous Al produced using small NaCl particles (30–50 μm) exhibited the trade-off relation between ΔP_{cap} and K via the porosity. This trend was consistent with the previous results of mono-porous Al fabricated using large NaCl particles (330–430 μm). The pores were well-connected to form large pores as the porosity increased, resulting in the trade-off relation.

(2) The mono-porous Al with a smaller pore size exhibited a higher ΔP_{cap} and a lower K compared to the mono-porous Al with a larger pore size under a constant porosity. This trend demonstrated the well-known trade-off relation between ΔP_{cap} and K via the pore size.

(3) The volume fractions of large and small pores in the bi-porous Al were successfully controlled under a constant total porosity of 70% by tailoring the blending volume fractions of large and small NaCl particles. Increasing the volume fraction of small pores from 0% to 30% increased the ΔP_{cap} from 197 to 381 Pa while slightly decreasing K from 10.4×10^{-11} m^2 to 8.4×10^{-11} m^2. When the volume fraction of small pores was increased to 40%, ΔP_{cap} and K degraded to 293 Pa and 7.4×10^{-11} m^2.

(4) Almost all the bi-porous samples exhibited intermediate ΔP_{cap} and K between the mono-porous samples with large and small pores alone. However, the optimized bi-porous structure with 40% large and 30% small pores exhibited a higher ΔP_{cap} and $\Delta P_{cap} \cdot K$ than the mono-porous samples. The bi-porous structure was not always superior to the mono-porous structure and must be controlled to improve ΔP_{cap} and $\Delta P_{cap} \cdot K$.

Supplementary Materials: The following supporting information can be downloaded at: https://www.mdpi.com/article/10.3390/ma17194729/s1, Figure S1: SEM images of feature pore structures of bi-porous Al: (a) L60S10, (b) L50S20, (c) L40S30, and (d) L30S40, Figure S2: Size distribution of small pores on cell walls of large pores: (a) L60S10, (b) L50S20, (c) L40S30, and (d) L30S40.

Author Contributions: Conceptualization, H.S. and A.S.; methodology, H.S. and A.S.; validation, A.S. and M.K.; formal analysis, H.S. and A.S.; investigation, H.S. and A.S.; resources, A.S. and M.K.; data curation, H.S. and A.S.; writing—original draft preparation, H.S. and A.S.; writing—review and editing, N.T., A.S. and M.K.; supervision, A.S.; project administration, M.K.; funding acquisition, M.K. All authors have read and agreed to the published version of the manuscript.

Funding: This research was supported by Grant-in-Aid for Challenging Research (Pioneering) (Grant number: 22548412) and JST ACT-X JST (ACT-X Grant Number JP 21460460), Japan.

Institutional Review Board Statement: Not applicable.

Informed Consent Statement: Not applicable.

Data Availability Statement: Data will be made available on request.

Conflicts of Interest: The authors declare no conflicts of interest.

References

1. Allioux, F.; Merhebi, S.; Tang, J.; Idrus-Saidi, S.A.; Abbasi, R.; Saborio, M.G.; Ghasemian, M.B.; Han, J.; Namivandi-Zangeneh, R.; O'Mullane, A.P.; et al. Catalytic Metal Foam by Chemical Melting and Sintering of Liquid Metal Nanoparticles. *Adv. Funct. Mater.* **2020**, *30*, 1907879. [CrossRef]
2. Malloy, J.; Quintana, A.; Jensen, C.J.; Liu, K. Efficient and Robust Metallic Nanowire Foams for Deep Submicrometer Particulate Filtration. *Nano Lett.* **2021**, *21*, 2968–2974. [CrossRef] [PubMed]
3. Yang, H.; Yang, Y.; Ma, B.; Zhu, Y. Experimental Study on Capillary Microflows in High Porosity Open-Cell Metal Foams. *Micromachines* **2022**, *13*, 2052. [CrossRef] [PubMed]
4. Xia, D.; Qin, Y.; Guo, H.; Wen, P.; Lin, H.; Voshage, M.; Schleifenbaum, J.H.; Cheng, Y.; Zheng, Y. Additively Manufactured Pure Zinc Porous Scaffolds for Critical-Sized Bone Defects of Rabbit Femur. *Bioact. Mater.* **2023**, *19*, 12–23. [CrossRef] [PubMed]
5. Chang, C.; Li, B.; Fu, B.; Yang, X.; Lou, T.; Ji, Y. Simulation and Experimental Investigation of Phase Change Materials within Porous Metal Structures with Gradient Porosity. *Int. J. Heat Mass Transf.* **2024**, *234*, 126136. [CrossRef]
6. Li, X.; Yao, D.; Zuo, K.; Xia, Y.; Zeng, Y.-P. Effects of Pore Structures on the Capillary and Thermal Performance of Porous Silicon Nitride as Novel Loop Heat Pipe Wicks. *Int. J. Heat Mass Transf.* **2021**, *169*, 120985. [CrossRef]
7. Nishikawara, M. Pore Network Simulation of Loop Heat Pipe Evaporator with Different Pore Size Distribution. *Int. J. Heat Mass Transf.* **2024**, *218*, 124739. [CrossRef]
8. Kumar, P.; Wangaskar, B.; Khandekar, S.; Balani, K. Thermal-Fluidic Transport Characteristics of Bi-Porous Wicks for Potential Loop Heat Pipe Systems. *Exp. Therm. Fluid Sci.* **2018**, *94*, 355–367. [CrossRef]
9. Feng, C.; Yugeswaran, S.; Chandra, S. Capillary Rise of Liquids in Thermally Sprayed Porous Copper Wicks. *Exp. Therm. Fluid Sci.* **2018**, *98*, 206–216. [CrossRef]
10. Liu, J.; Zhang, Y.; Feng, C.; Liu, L.; Luan, T. Study of Copper Chemical-Plating Modified Polyacrylonitrile-Based Carbon Fiber Wick Applied to Compact Loop Heat Pipe. *Exp. Therm. Fluid Sci.* **2019**, *100*, 104–113. [CrossRef]
11. Zhang, J.; Lian, L.; Liu, Y.; Wang, R. The Heat Transfer Capability Prediction of Heat Pipes Based on Capillary Rise Test of Wicks. *Int. J. Heat Mass Transf.* **2021**, *164*, 120536. [CrossRef]
12. Lu, L.; Sun, J.; Liu, Q.; Liu, X.; Tang, Y. Influence of Electrochemical Deposition Parameters on Capillary Performance of a Rectangular Grooved Wick with a Porous Layer. *Int. J. Heat Mass Transf.* **2017**, *109*, 737–745. [CrossRef]
13. Li, J.; Zhang, M. Biporous Nanocarbon Foams and the Effect of the Structure on the Capillary Performance. *Prog. Nat. Sci.* **2020**, *30*, 360–365. [CrossRef]
14. Zhang, S.; Chen, C.; Chen, G.; Sun, Y.; Tang, Y.; Wang, Z. Capillary Performance Characterization of Porous Sintered Stainless Steel Powder Wicks for Stainless Steel Heat Pipes. *Int. Commun. Heat Mass Transf.* **2020**, *116*, 104702. [CrossRef]

15. Li, X.; Yao, D.; Zuo, K.; Xia, Y.; Yin, J.; Liang, H.; Zeng, Y.-P. Fabrication, Microstructural Characterization and Gas Permeability Behavior of Porous Silicon Nitride Ceramics with Controllable Pore Structures. *J. Eur. Ceram. Soc.* **2019**, *39*, 2855–2861. [CrossRef]
16. Mottet, L.; Prat, M. Numerical Simulation of Heat and Mass Transfer in Bidispersed Capillary Structures: Application to the Evaporator of a Loop Heat Pipe. *Appl. Therm. Eng.* **2016**, *102*, 770–784. [CrossRef]
17. Wang, D.; Wang, J.; Bao, X.; Chen, G.; Chu, H. Experimental Study on Hydraulic and Thermal Characteristics of Composite Porous Wick with Spherical–Dendritic Powders. *J. Therm. Anal. Calorim.* **2020**, *141*, 107–117. [CrossRef]
18. Byon, C.; Kim, S.J. Capillary Performance of Bi-Porous Sintered Metal Wicks. *Int. J. Heat Mass Transf.* **2012**, *55*, 4096–4103. [CrossRef]
19. Cao, Y.; Wu, D.; Guo, C.; Zou, Y. Fabrication and Capillary Performance of Bi-Porous Ti3AlC2 Wicks with Controllable Pore Size Proportion Using Dissolvable Pore Formers. *J. Mater. Res. Technol.* **2021**, *15*, 4370–4380. [CrossRef]
20. Cao, Y.; Guo, C.; Wu, D.; Zou, Y. Effects of Pore Structure Characteristics on Performance of Sintered Bi-Porous Ti3AlC2 Wicks. *Mater. Res. Express* **2021**, *8*, 015602. [CrossRef]
21. Zhu, P.; Wu, Z.; Zhao, Y. Hierarchical Porous Cu with High Surface Area and Fluid Permeability. *Scr. Mater.* **2019**, *172*, 119–124. [CrossRef]
22. Li, H.; Fu, S.; Li, G.; Fu, T.; Zhou, R.; Tang, Y.; Tang, B.; Deng, Y.; Zhou, G. Effect of Fabrication Parameters on Capillary Pumping Performance of Multi-Scale Composite Porous Wicks for Loop Heat Pipe. *Appl. Therm. Eng.* **2018**, *143*, 621–629. [CrossRef]
23. Shu, Y.; Suzuki, A.; Takata, N.; Kobashi, M. Fabrication of Porous NiAl Intermetallic Compounds with a Hierarchical Open-Cell Structure by Combustion Synthesis Reaction and Space Holder Method. *J. Mater. Process. Technol.* **2019**, *264*, 182–189. [CrossRef]
24. Shu, Y.; Suzuki, A.; Takata, N.; Kobashi, M. Effect of Powder Blending Ratio on Porous Structure Formed by Combustion Synthesis Reaction between Ni and Al Together with Space Holder. *J. Funct. Graded Mater.* **2019**, *33*, 38–43. [CrossRef]
25. Kobashi, M.; Inoguchi, N.; Kanetake, N. Effect of Elemental Powder Blending Ratio on Combustion Foaming Behavior of Porous Al–Ti Intermetallics and Al3Ti/Al Composites. *Intermetallics* **2010**, *18*, 1039–1045. [CrossRef]
26. Zhao, Y.Y.; Sun, D.X. A Novel Sintering-Dissolution Process for Manufacturing Al Foams. *Scr. Mater.* **2001**, *44*, 105–110. [CrossRef]
27. Shen, H.; Suzuki, A.; Takata, N.; Kobashi, M. Elucidating Dominant Flow Channel Size for Capillary Performance of Open-Cell Porous Wicks. *Int. J. Heat Mass Transf.* **2024**, *223*, 125217. [CrossRef]
28. Suzuki, A.; Arai, Y.; Takata, N.; Kobashi, M. Structural Design and Bonding Strength Evaluation of Al/Epoxy Resin Joint via Interpenetrating Phase Layer. *J. Mater. Process. Technol.* **2018**, *262*, 11–18. [CrossRef]
29. Tang, Y.; Xi, X.; Liang, F.; Zhang, S.; Tang, H.; Wu, C.; Sun, Y. Enhanced Capillary Performance of Ultrathin Nylon Mesh Wick for Flexible Thermal Management Systems. *Int. J. Heat Mass Transf.* **2023**, *200*, 123545. [CrossRef]
30. Kim, S.-G. Development of Direct Joining Process of Metals and Polymers via Additively Fabricated Anchor Layer by Laser Induced In-Situ Reaction. Ph.D. Thesis, Nagoya University, Nagoya, Japan, 2020.

materials

MDPI

Article

Numerical Study on Fluid Flow Behavior and Heat Transfer Performance of Porous Media Manufactured by a Space Holder Method

Xianke Lu [1], Yuyuan Zhao [1,2,*], Yue Zhang [3] and Mian Wu [1]

1 School of Mechanical and Automotive Engineering, Ningbo University of Technology, Ningbo 315211, China; xianke@nbut.edu.cn (X.L.); wumian91@nbut.edu.cn (M.W.)
2 School of Engineering, University of Liverpool, Liverpool L69 3GH, UK
3 China Merchants Chongqing Communications Research & Design Institute Co., Ltd., Chongqing 400067, China; zhangyue8@cmhk.com
* Correspondence: zhaoyuyuan@nbut.edu.cn

Abstract: The velocity field and temperature field are crucial for metal foams to be used as a heat exchanger, but they are difficult to obtain through physical experiments. In this work, the fluid flow behavior and heat transfer performance in open-cell metal foam were numerically studied. Porous 3D models with different porosities (55–75%) and pore sizes (250 μm, 550 μm, and 1000 μm) were created based on the porous structure manufactured by the Lost Carbonate Sintering method. A wide flow velocity range from 0.0001 m/s to 0.3 m/s, covering both laminar and turbulent flow regimes, is fully studied for the first time. Pressure drop, heat transfer coefficient, permeability, form drag coefficient, temperature and velocity distributions were calculated. The calculated results agree well with our previous experimental results, indicating that the model works well. The results showed that pressure drop increased with decreasing porosity and increasing pore size. Permeability increased and the form drag coefficient decreased with increasing porosity, and both increased with increasing pore size. The heat transfer coefficient increased with increasing velocity and porosity, whereas it slightly decreased with increasing pore size. The results also showed that at high velocity, only the metal foam close to the heat source contributes to heat dissipation.

Keywords: open-cell metal foam; porous flow; pressure drop; heat transfer coefficient; forced convection; CFD

Citation: Lu, X.; Zhao, Y.; Zhang, Y.; Wu, M. Numerical Study on Fluid Flow Behavior and Heat Transfer Performance of Porous Media Manufactured by a Space Holder Method. *Materials* **2024**, *17*, 2695. https://doi.org/10.3390/ma17112695

Academic Editor: Jaroslav Kovacik

Received: 1 April 2024
Revised: 20 May 2024
Accepted: 27 May 2024
Published: 3 June 2024

1. Introduction

Thermal management has always been an important issue in the electronic industry since an overly high temperature will not only reduce the performance of electronic components, but also shorten their service life [1,2]. With the development of electric vehicles, thermal management becomes more important because there is a recommended operating temperature for the battery and this temperature is related to the performance and safety of the vehicles [3,4]. Traditional designs, such as louvered fins and slit fins, are hitting bottlenecks due to their limits in mass and specific surface area. Open-cell metal foams as an alternative for the traditional fins are attracting more and more attentions [5–7].

An open-cell metal foam is a cellular structure consisting of a network of metal ligaments (copper and aluminum) and fluid-filled (usually water and air) pores. The ratio of pore volume to the total volume is named porosity with a value normally ranging from 50% to 95%. The metal network provides a low density, good thermal conductivity, and an excellent fluid mixing function. The specific surface area of metal foam reaches 800 cm^2/g, and combined with the good fluid permeability, it provides favorable conditions for heat convection [8]. These properties make open-cell metal foams widely used in the field of heat transfer, chemical catalysis and filtration [9].

Open-cell metal foams can be manufactured by investment casting, powder metallurgy sintering, and vapor deposition process methods [10,11]. Open-cell copper foam manufactured by the Lost Carbonate Sintering (LCS) process which is based on the powder metallurgy sintering technique is a promising type of material for use as a heat exchanger [12]. These foams have two grades of pores that provide more surface area for heat transfer than other cast foams. In our previous research, the heat transfer performance and fluid flow behavior in these LCS foams were experimentally investigated by the pressure drop method and flow visualization method [13,14]. The effects of flow rate and flow regime on pressure drop and heat transfer were systematically analyzed, which were generally consistent with other cellular materials. However, some challenges are difficult to overcome in physical experiments, such as low flow control, low pressure, and low velocity measurements and temperature distribution throughout the metal foam. Numerical simulation methods are powerful tools for studying coupled flow and heat transfer in metal foam, providing more information than physical experiments [15–17].

Model construction is the most essential and important step in numerical simulation. Metal foams are complex cellular structures and it is impossible to precisely describe their structure characteristics. In addition, it is unnecessary to use a complex model when we focus on the fundamentals. Therefore, a simplified structure analogous to the metal foam is preferred in research. Lu et al. [18] studied the effects of foam density, pore size, and inlet length on the temperature distribution by using simple cubic unit cells with a constant wall temperature setup and a discrepancy was found due to oversimplification of the transport equation. Boomsma et al. [19] defined an idealized open-call model using eight 14-sided tetrakaidecahedron cells (Kelvin model) and used periodic boundary conditions to model an infinity matrix. They found the simulated pressure drop values were 25% lower than the experimental values due to the lack of wall effects. Kopanidis et al. [20] built a solid ligament foam model based on a volume composed of six tetrakaidecahedra and two irregular dodecahedra (W-P model). This model results in a 0.3% reduction in surface energy compared to the Kelvin model used by Boomsma. Results showed that the heat transfer and pressure drop significantly depend on entrance effects, near-wall effects and heat conduction. Recently, Hu et al. [21] used the Kelvin model to study the forced convection of porous metal used as a heat exchanger in a turbine engine at high velocity (>20 m/s) which is difficult to achieve in experiment. A quadratic relationship between pressure drop and velocity, and a logarithmic relationship between the volumetric heat transfer coefficient and velocity were confirmed at high velocity. Yang et al. [22] investigated the interstitial heat transfer coefficient of high-porosity foam using the W-P model under rotational conditions. A relationship consistent with the experimental results was established between the heat transfer coefficient and the structural parameters, and it was also found that the interstitial heat transfer coefficient increases with increasing flow rate and rotation speed. Gauna and Zhao [23] studied pressure drop and heat transfer in porous media but did not consider the effect of natural convection on heat transfer.

However, the difference between the currently existing models and the LCS structure is so great that the current model cannot be used directly. In this study, a geometry model is proposed to imitate the LCS porous structure. The design process of the model, combined with the fabrication process of the LCS metal foam, is fully described. The pressure drop and heat transfer coefficient are systematically investigated for models with different porosities, pore sizes and flow rates. Temperature distribution and velocity distribution are analyzed at various parameters. This study can provide more information for the structure design of metal foams used as heat exchangers.

2. Numerical Model and Method

2.1. Model Construction

The present model is based on an open-cell metal foam manufactured by the space holder method [11]. In manufacturing, large carbonate particles and fine metal powder are mixed, compacted and sintered. Carbonate particles are removed by a dissolution or decomposition process, leaving an open-cell metal foam. The pore distribution and connectivity of the pores are random on a macroscopic scale, but regularity can be found at the microscopic scale. A schematic drawing of the microstructure and scanning electron micrograph of copper foam sample manufactured by this method are shown in Figure 1. The large pores are the replica of the carbonate particles and these pores connect to each other through a small hole which is formed by the tangent point. The pore wall is composed of sintered copper particles. The porosity of the metal foam manufactured by this process ranges from 50% to 80%. The median porosity of such foam makes the Kelvin model [24] infeasible as it can only show the structure characteristics of the high-porosity foam. The face-centered cubic (FCC) packed unit cell has a packing factor of 74%, which matches the current porosity very well. All models in the present study are constructed based on the FCC unit cell.

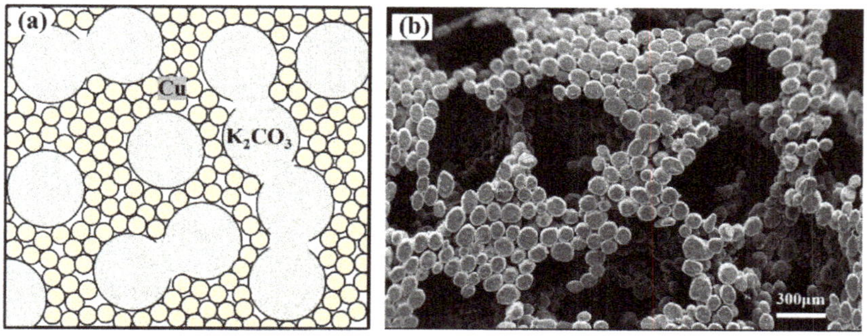

Figure 1. Structure of porous copper sample manufactured by a space holder method: (**a**) schematic diagram of the mixture of pore agent and metal powder and (**b**) scanning electron micrograph of sintered sample.

2.2. Physical Model

The representative elementary volume (REV) is used to represent the porous structure of metal foam. Figure 2 shows the design procedure of the REV unit cell. The unit cell of the REV is the negative replica of the FCC unit cell in which the spherical pores are located where the atoms are in the FCC unit cell. The pores are connected through cylindrical holes at the tangent point. The porosity of the REV is controlled by varying the distance of the neighboring pores and the diameter of the cylindrical hole. Figure 3 shows the geometric relationship of the cylindrical hole and neighboring pore and particles. The radius of the cylindrical hole, r_c, reflects the connectivity of the neighboring pores and can be determined by the pore radius, R, and metal particle radius, r. The distance between pores can be calculated by Equation (1) [23]:

$$l_p = 2R\left(\sqrt[3]{\frac{\pi}{\left(3\sqrt{2}\right)\epsilon}} - 1\right) \quad (1)$$

The radius of the cylindrical hole is calculated by Equation (2) [25]:

$$r_c = \sqrt{\frac{\omega \cdot A_{SC}}{12\pi}} \quad (2)$$

where ω is the coordination number, representing the number of connected pores with its neighbors. The coordination number is 12 for the FCC, but for the foam manufactured by a space holder method, the value is much lower and can be estimated by Equation (3) [25]:

$$\omega = \frac{2}{\left(1 - \sqrt{\frac{\phi+2}{\sqrt{\phi^2+6\phi+5}}}\right)\left(1 - \phi + \frac{\phi}{\epsilon}\right)} \tag{3}$$

where ϕ is radius ratio between the pore and metal particle. A_{SC} is the area of the sphere crown and can be calculated by Equation (4) [25]:

$$A_{SC} = 2\pi R^2\left(-\frac{\phi+2}{\sqrt{\phi^2+6\phi+5}}\right) \tag{4}$$

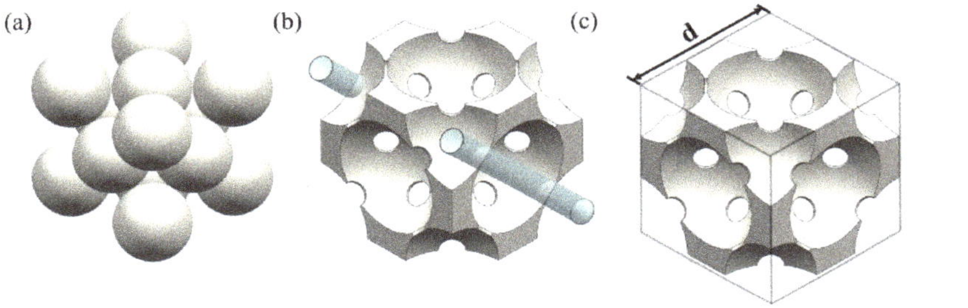

Figure 2. The design of the REV unit cell. From (**a**) FCC packing to (**b**) negative replicas and then to (**c**) final unit cell.

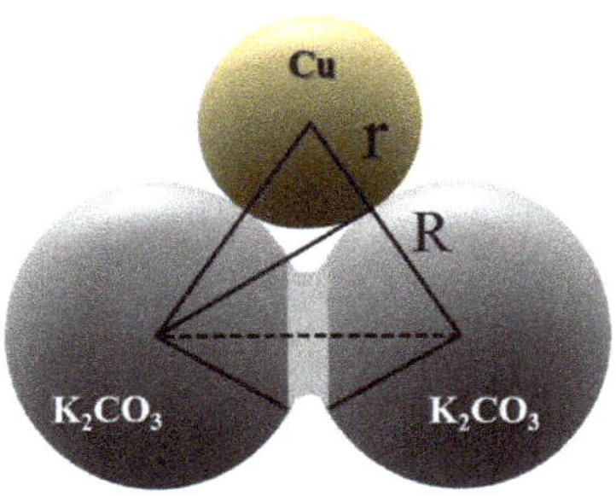

Figure 3. Schematic drawing showing the geometric relationship of the cylindrical hole and neighboring particles.

Models with pore sizes of 250 μm, 550 μm and 1000 μm and porosities of 55%, 60%, 65%, 70% and 75% were built. In this study, the diameter of the cylindrical hole was calculated assuming that the diameter of the metal particle is 50 μm. The details of all the physical models are listed in Table 1. Figure 4 shows the typical geometry model and boundary conditions of simulation. Table 2 lists details of the boundary condition. According to the previous study, the result is unchanged when the number of unit cells in the mean flow direction is equal to or larger than four [23]. Therefore, five cells were built in the mean flow direction. The thickness (in heat flow direction) of the sample significantly affects the heat transfer performance of the metal foam. To investigate the effect of thickness, models with 1–4 layers at porosities of 55% and 75% and a pore size of 550 μm were built. For other results, models with 3 layers were used. The dimensions of a typical three-layered model are around 4.5 × 0.9 × 2.7 mm^3. A total of 21 models were built and more than 400 calculations were performed.

Table 1. Diameter of the cylindrical hole and neighboring pore distance for all physical models.

Porosity	55%		60%		65%		70%		75%	
	l_p	r_c	l_p	r_c	l_p	r_c	l_p	r_c	l_p	r_c
250 μm	26.0	32.0	18.2	34.7	11.1	37.6	4.7	40.7	1.1	44.2
550 μm	57.3	50.2	39.9	55.0	24.4	60.3	10.4	66.4	2.3	73.5
1000 μm	104.2	69.3	72.6	76.2	44.4	84.1	18.9	93.3	42.6	104.3

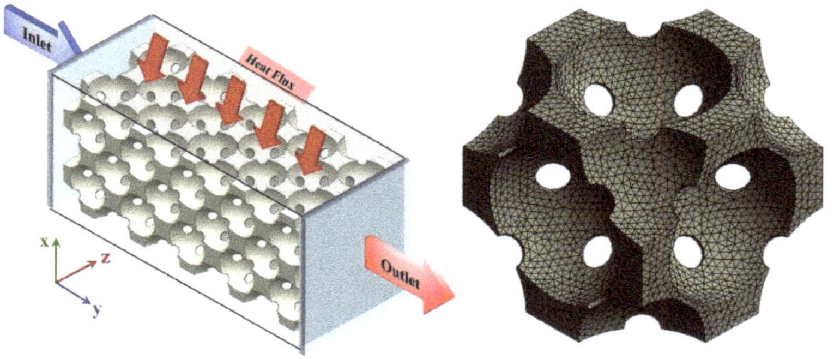

(a) (b)

Figure 4. Computational domain and boundary conditions of the (**a**) geometry and (**b**) grid of the unit cell.

Table 2. Boundary condition settings.

Boundary Name	Boundary Condition
Top surface	Constant heat flux
Front surface	Symmetric
Back surface	Symmetric
Bottom surface	Zero heat flux

2.3. Governing Equation and Boundary Condition

Heat transfer and fluid flow were simulated by ANSYS Fluent [26]. The working fluid velocity ranges from 0.0001 to 0.3 m/s, which covers both laminar and turbulent. A heating block was placed on the top of the model so that the natural convection within the porous model was avoided. The highest temperature of the heating block and the working fluid is less than 100 °C. Radiation is negligible and the thermo-physical properties such as thermal conductivity are temperature independent [27]. In this study, water was used as the working fluid and it was assumed to be incompressible. The temperature field of the porous model is calculated by solving the conjugate heat transfer problem. In the case of the porous model, conjugate heat transfer refers to the simultaneous heat transfer through both the solid material of the porous medium and the fluid flowing through the pores. The conjugate heat transfer problem is typically solved using numerical methods such as finite element analysis or computational fluid dynamics. These methods allow for the simulation of heat transfer within complex geometries and can account for different modes of heat transfer such as conduction, convection, and radiation. The temperature field in the porous medium is influenced by the heat transfer between the solid matrix and the fluid, as well as by any heat sources or sinks present in the system. Solving the conjugate heat transfer problem is crucial for applications such as heat exchangers, thermal insulation or energy storage systems where efficient heat transfer is essential.

The incompressible working fluid is governed by the Navier–Stokes equations, which includes the continuity, momentum and energy equation given by Equations (5)–(7).

$$\text{Continuity } \frac{\partial}{\partial x_i}\rho u_i = 0 \tag{5}$$

$$\text{Momentum } \frac{\partial}{\partial x_i}(\rho u_i u_j) = \frac{\partial P}{\partial x_j} + \frac{\partial}{\partial x_i}\left(\mu \frac{\partial u_j}{\partial x_i}\right) \tag{6}$$

$$\text{Energy } \frac{\partial}{\partial x_i}(\rho u_i C_p T) = \frac{\partial}{\partial x_i}\left(k \frac{\partial T}{\partial x_i}\right) \tag{7}$$

where ρ is the fluid density, u is the velocity, P is the pressure, μ is the fluid viscosity, T is the temperature and k is the thermal conductivity.

For laminar flow, the laminar model was used, while for turbulent flow, the standard k-ξ model was used. The critical velocity for the transition from laminar flow to turbulent flow was chosen according to both the experimental and simulation results. Our previous experiments [14] on porous copper reveal that the transition velocity is around 0.03 m/s. Simulation results shows that when the inlet velocity is higher than 0.04 m/s, the laminar model cannot make the simulation converge. According to the ANSYS documentation, if a simulation is set up using laminar flow, but the real flow is turbulent, convergence is difficult and the result is not correct. In the present study, simulation results agree well with the experimental results and therefore the transition velocity is around 0.04 m/s.

As shown in Figure 4, the computational domain is composed of an inlet region, porous model and outlet region. The inlet region was left for long enough that the flow was fully developed before entering the porous model [13]. A constant temperature of 300 K was set in the inlet boundary; a constant zero pressure was set in the outlet boundary; a constant heat flux of 250 KW/m^2 was set on the top of the heating block; and the interface between the metal and fluid was set as a non-slip boundary condition and the other two sides of the domain were set as symmetric. In this case, the heat is transferred from the heating block to the metal foam through conduction and then to the working fluid through convection. The temperature and velocity fields in the fluid domain were calculated and the temperature field in the solid domain was calculated.

2.4. Mesh and Solution

The 3D model was constructed using Unigraphics NX (UG) software (https://plm.sw.siemens.com/en-US/nx/) and then imported into the Integrated Computer Engineering and Manufacturing (ICEM) software (https://www.ansys.com/training-center/course-catalog/fluids/introduction-to-ansys-icem-cfd) to generate mesh. Different meshing methods and accuracies were applied in different regions to save calculating time. Coarse mesh was generated in the inlet, outlet and heating block region and fine mesh was generated in the porous model region. We also conducted a mesh number dependence study, and the results showed that when the number of meshes was higher than 6 million, the simulation results became stable. Considering accuracy and efficiency, approximately 10 million meshes were used in a single model. The overall quality of the mesh was higher than 0.85. The results were considered to be converged when the residuals of the variables were lowered by six orders of magnitude. The raw data were exported to files and then imported into MATLAB (https://www.mathworks.com/products/matlab.html) for post-processing.

2.5. Pressure Drop

It is difficult to choose two points in a porous medium to directly calculate the pressure drop due to the large difference in pressure in different regions of the porous medium. Therefore, the pressure drop was calculated from the pressure outside the porous model. When calculating the pressure drop, two planes were selected. One is located just in front of the inlet section and the other is located just after the outlet section. The average pressure on the two planes was calculated separately and the pressure drop value is equal to the pressure difference between the two planes. The same measurement method was used in our previous experimental study, and the consistency of the method facilitates data comparison.

2.6. Heat Transfer Coefficient

The overall heat transfer coefficient of the porous model was calculated according to the Newton's cooling law given by Equation (8)

$$J = h(T_c - T_{in}) \tag{8}$$

where J is the input heat flux, h is the heat transfer coefficient, T_c is the average temperature of the contact region between heating block and porous media and T_{in} is the temperature of the water in the inlet region.

3. Results and Discussion

3.1. Pressure Drop

Figure 5 shows the variation in pressure drop with velocity for models with different layers, porosities and pore sizes. To verify the effect of the number of layers (FCC unit cell) on the pressure drop, models with 1–4 layers were built, and the pressure drop at different velocities was calculated for each model. Figure 5a shows that the pressure drop is slightly higher for the one-layer model and the other models have the same pressure drop. Therefore, the three-layer model was used for all simulations. Figure 5b shows that the pressure drop increases dramatically with velocity and a typical pressure drop–velocity curve was achieved, which is consistent with previous physical experiments [28]. The pressure drop for the experimental sample with a porosity of 59% is higher than that for the model with a porosity of 55%. This is because the pore connectivity of the copper foam sintered by the space holder method is poor, while the pore connectivity of the model is not affected by the porosity. It also shows that the rate at which the pressure drop increases with velocity is much higher for the low-porosity model than for the high-porosity model.

The relationship between pressure drop and velocity depends on the flow regime. At low velocity, viscous force dominates the flow and a linear relationship can be observed. This linear relationship can be described by Darcy's law [29]: $\frac{\Delta P}{L} = \frac{\mu V}{K}$, where ΔP is pressure drop, L is the length of the porous media, V is the Darcian velocity which equals to the volume flow rate divided by the cross-sectional area of the flow channel and K is the permeability of the porous medium. The flow regime that applies Darcy's law is called the Darcy regime. Usually, the Darcy regime is very narrow compared to other regimes.

At high velocity, the inertial force plays a significant role and dominates the flow. The Forchheimer equation is commonly used to describe the relationship between pressure drop and velocity. The equation is in the form of:

$$\frac{\Delta P}{L} = \frac{\mu}{K}V + \rho C V^2 \tag{9}$$

where C is the form drag coefficient. The Forchheimer equation is a quadratic function and applies the non-Darcy flow regime. Fitting curves based on quadratic function are plotted in the Figure 5b and a high degree of consistency is observed.

Figure 5b,c show that both porosity and pore size have a significant effect on the pressure drop. For the model with a pore size of 550 μm, the pressure drop for porosity 55% is nearly eight times that of porosity 75% at 0.25 m/s. For a specific porosity and velocity, e.g., 65% and 0.25 m/s, the pressure drop for pore size 1000 μm is 2–3 times that of pore size 250 μm. For a specific porosity, a large pore size means a low number of pores, resulting in high tortuosity, while a small pore size means a high number of pores, resulting in low tortuosity. The tortuosity has a great influence on the pressure drop. For porous media with high tortuosity, the fluid has to travel a longer distance from point A to point B, and the pressure drop will be higher [30].

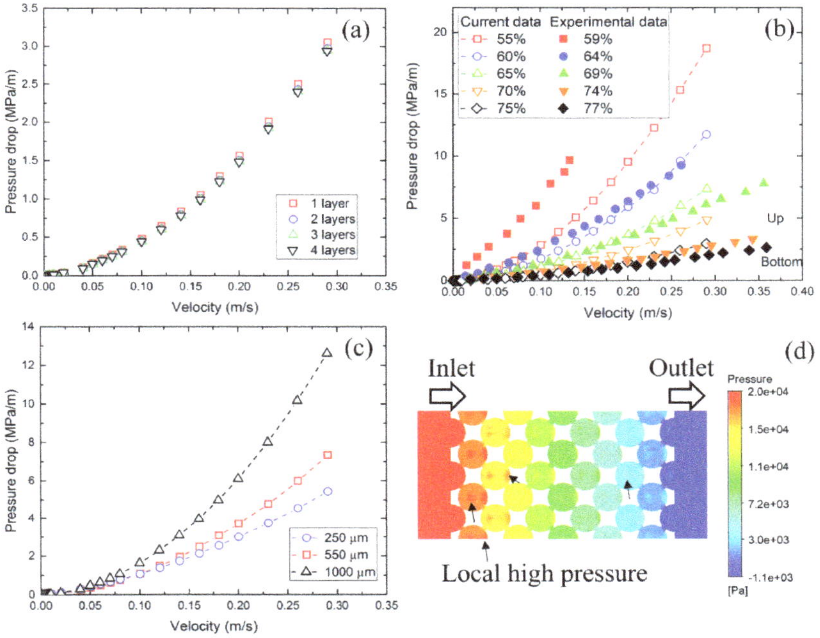

Figure 5. (**a**) Effect of the number of layers on pressure drop; (**b**) variation in pressure drop with velocity for different models with pore size of 550 μm and comparison with experimental data [28]; (**c**) pressure drop comparison of models with porosity of 65% and different pore sizes; and (**d**) pressure field in a model with porosity 65%, pore size 550 μm and velocity 0.2 m/s.

3.2. Permeability and Friction Factor

Permeability and the inertial drag coefficient were calculated by comparing the fitting equation with the Forchheimer equation. The values of permeability (K) and the inertial drag coefficient are listed in Table 3. The variation in permeability and the inertial drag coefficient with porosity and pore size are shown in Figure 6. As porosity increases, the permeability increases and the form drag coefficient decreases. The effect of pore size on permeability and the form drag coefficient is related to porosity. The effect of pore size on permeability is much stronger at high porosity than at low porosity. For the form drag coefficient, the opposite is true.

Table 3. Permeability (K) and inertial drag coefficient (C) of all models.

	Pore Size					
	250 μm		550 μm		1000 μm	
Porosity	$K \times 10^{-7}$ (m^2)	C (m^{-1})	$K \times 10^{-7}$ (m^2)	C (m^{-1})	$K \times 10^{-7}$ (m^2)	C (m^{-1})
55%	0.38	197.3	0.25	816.6	0.19	3000.0
60%	0.56	124.5	0.44	499.3	0.38	2000.0
65%	0.83	78.2	0.74	308.2	0.76	1000.0
70%	1.16	49.9	1.14	198.7	1.83	663.5
75%	1.49	32.1	1.82	117.9	2.75	397.9

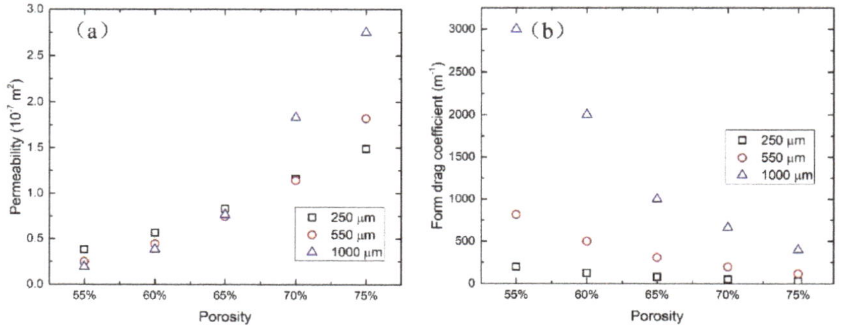

Figure 6. The variation of (**a**) permeability and (**b**) form drag coefficient with porosity and pore size.

The flow regimes cannot be identified from the pressure drop–velocity figure but can be figured out from the relationship between friction factor and velocity. The friction factor is defined as:

$$f = \frac{\Delta P}{L} \cdot \frac{D}{\rho V^2} \tag{10}$$

Figure 7 shows the variation in friction factor with velocity. The linear relationship between friction factor and velocity at low velocity corresponds to the laminar flow regime. A constant friction value will be reached at high velocity, which corresponds to the turbulent flow regime. A transition range exists between laminar and turbulent flow regimes. Figure 7 shows that high porosity and a large pore size lead to low friction because of weak impedance of the solid matrix and a low surface area. A large pore size also makes the flow become stable turbulent faster.

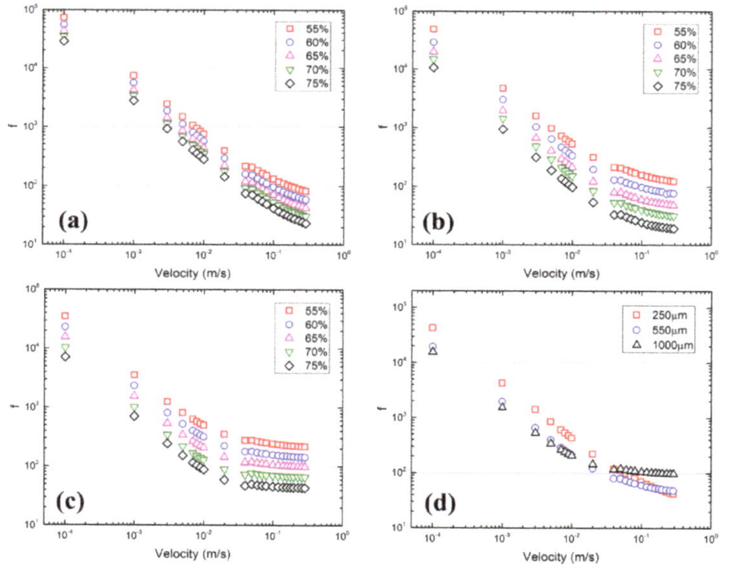

Figure 7. The friction factor. (**a**) 250 μm, (**b**) 550 μm, (**c**) 1000 μm and (**d**) porosity is 65%.

3.3. Temperature Distribution

3.3.1. Effect of Flow Rate

Figure 8 shows the effect of flow rate on the temperature distribution in metal foam with a pore size of 550 μm and a porosity of 65%. At different velocities, the difference in the highest temperature in the metal foam can reach up to 40 K. For example, the highest temperature is more than 340 K at a velocity of 0.003 m/s, while it is only about 300 K at a velocity of 0.29 m/s. The uniformity of temperature distribution throughout the whole metal foam also varies greatly at different velocities. When the velocity is less than 0.02 m/s, the heat reaches the bottom part (away from the heat source) of the metal foam, while when the velocity is higher than 0.02 m/s, the bottom is not heated at all, which means that at high velocity, the bottom of the metal foam contributes nearly nothing to heat transfer because the heat has been taken away before it gets that far.

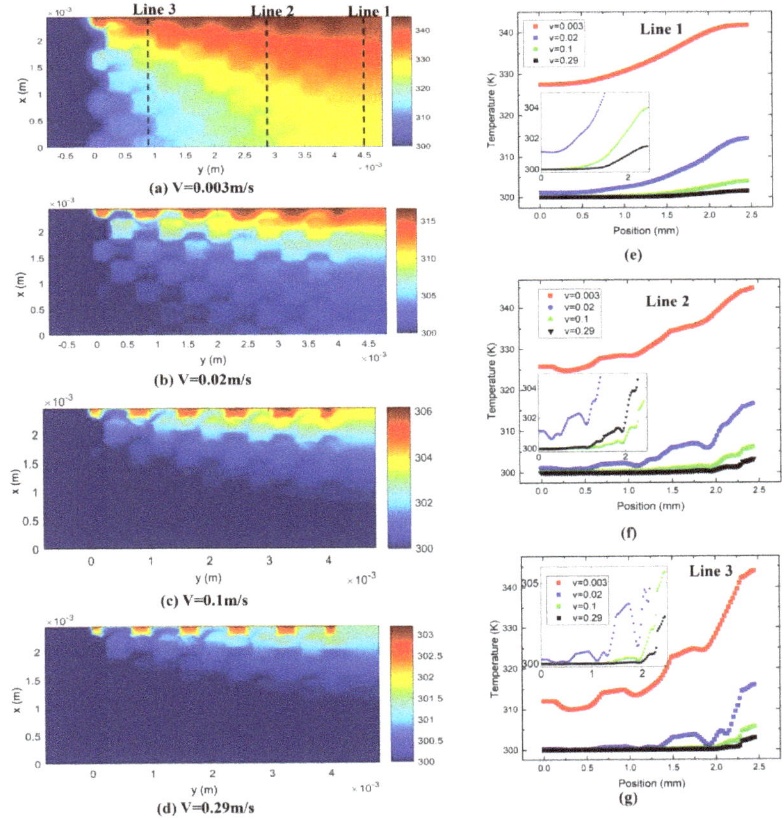

Figure 8. (**a**–**d**) Temperature distribution in metal foam under different flow rates. Heat source is on the top of the wall. Temperature distribution along the (**e**) line 1, (**f**) line 2 and (**g**) line 3 in (**a**–**d**).

Figure 8e–g show the temperature distribution along the three reference lines in Figure 8a. Line 2 and line 3 are located on the metal foam, while line 1 is located in the outlet area. Figure 8f,g show that the temperature in the metal foam is higher than that of the fluid even though they are in direct contact. This is more pronounced in the area near the inlet because heat conduction in the metal foam happens faster than convection between the metal foam and the fluid.

3.3.2. Effect of Porosity

Figure 9 shows the temperature distribution along line 1 and 3 for samples with a porosity of 55%, 65% and 75% at low and high flow rates. Porosity has a great impact on the temperature distribution and this effect is more pronounced at a low flow rate near the inlet (line 3) and outlet (line 1). For example, in Figure 9b, the temperature difference between the heat source and cold wall is 60 K for the sample with a porosity of 75%, while the difference is only 15 K for the sample with a porosity of 55%. This is because, at low velocity (v = 0.003 m/s), the fluid cannot effectively take heat away. Therefore, if the heat cannot be transferred quickly to the cold part on the metal foam, there will be a large temperature difference between the heat source part and the cold wall part, which is the case for the high-porosity sample. However, if the heat can be efficiently transferred to the other side of sample due to high thermal conductivity, the temperature will be more uniform, which is the case for the low-porosity sample.

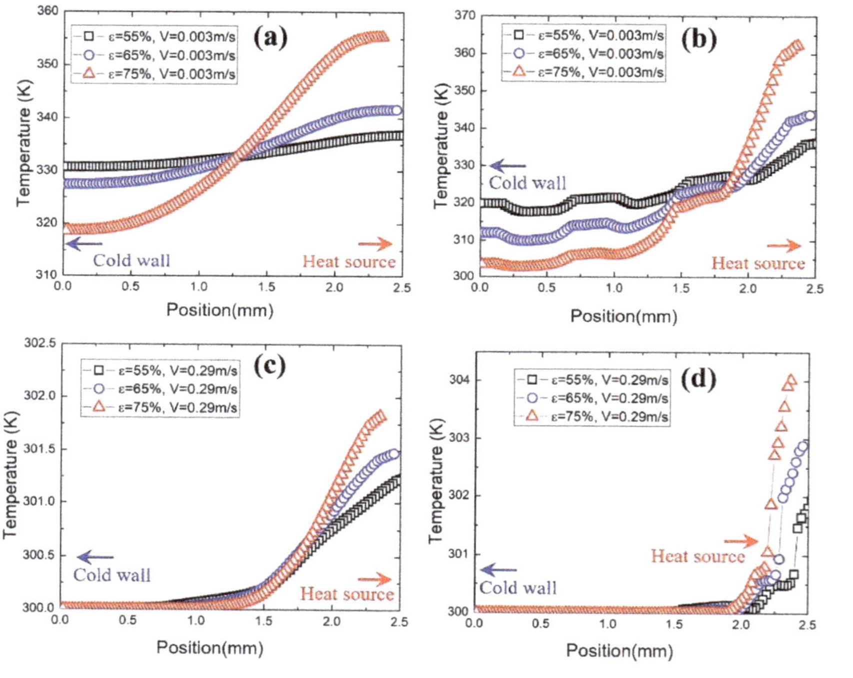

Figure 9. Temperature distribution along line 1 and 3 for samples with different porosities at different flow rates. The pore size of all samples is 550 µm. (**a**,**c**) show the temperature along line 1 and (**b**,**d**) show the temperature along line 3.

At high velocity (v = 0.29 m/s), the effect of porosity on the temperature distribution is also affected by the velocity, especially near the cold wall part. This is because the fast-moving fluid takes heat away before it is transferred to the cold wall part. Therefore, for samples with different porosities, the temperature near the cold wall part is the same. In addition, due to the strong influence of the high flow rate, the temperature difference in the heat source part of different samples is only 1–2 K.

3.3.3. Effect of Pore Size

Although all models have the same number of cell layers in the Z direction, a different pore size means different unit cell dimensions and different model heights. Figure 10 shows the effect of pore size on the temperature distribution along line 2 for metal foam with different pore sizes. In regions close to the heat source, the larger the pore size, the higher the temperature for all velocities. However, in the cold wall part, the smaller the pore size, the higher the temperature at low velocity, and the same temperature at high velocity. It has been proved that the thermal conductivity of metal foam depends insignificantly on pore size [31–33], so the temperature difference on the metal foam is mainly caused by the height difference of the model.

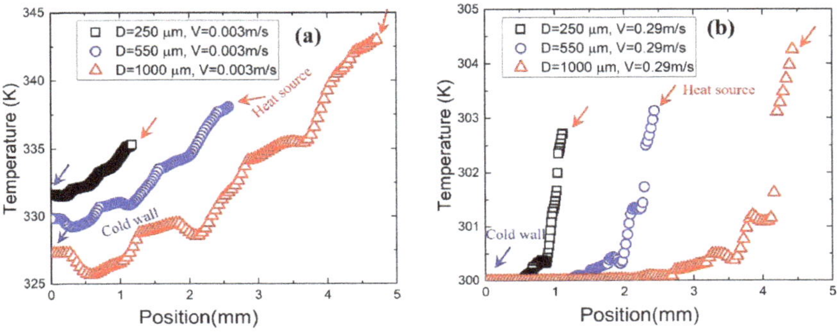

Figure 10. Temperature distribution along line 2 for metal foam with porosity of 55% and pore sizes of 250 μm, 550 μm and 1000 μm at velocity of (**a**) 0.003 m/s and (**b**) 0.29 m/s.

3.4. Velocity Distribution

3.4.1. Effect of Flow Rate

Figure 11 shows the x–y plane of the pore-scale velocity distribution of a sample with a pore size of 550 μm and a porosity of 65% at different flow rates. When the flow velocity is less than 0.02 m/s, the direction of the mean velocity at the pore scale is inlet to outlet, while when the flow velocity is great than 0.1 m/s, vortices can be found in the top, bottom and left part of a pore. As discussed in the Figure 8, a fast-moving (v > 0.1 m/s) fluid has efficiently removed the heat from the metal matrix, making the lower part of the metal foam approximately the same temperature as water. This greatly enhanced convective heat transfer capacity is mainly attributed to pore-scale vortices. The high temperature near the heat source region proves that heat conduction through the metal foam is still faster than heat convection.

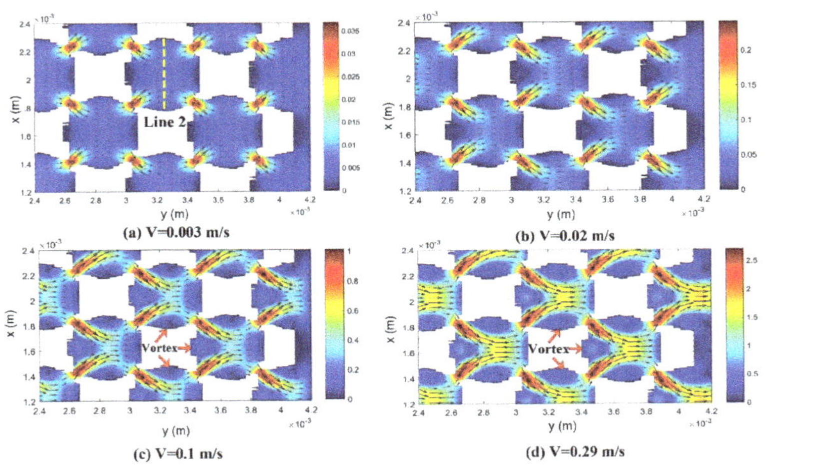

Figure 11. Pore-scale velocity distribution at different flow rates. Pore size: 550 μm; porosity: 65%.

Figure 12 shows the velocity profiles along a reference line at different flow rates for the sample with a pore size of 550 μm and a porosity of 65%. Line 1 is located at the outlet region without the metal foam. When the velocity is less than or equal to 0.1 m/s, the velocity profile in the non-near-wall region is quite flat. This is due to the good mixing function of metal foam. Meanwhile, when the velocity reaches 0.29 m/s, a "W"-shaped profile is observed in the non-near-wall region. This demonstrates that a longer distance after the outlet is required at high flow rates to obtain a uniformly distributed velocity. Line 2 is located at the position marked in Figure 11a and its velocity profile is shown in Figure 12b. A typical "M" shape is observed due to the merging of two tributaries in the pore. However, due to the influence of eddy flow, the velocity near the pore wall when the flow rate is 0.29 m/s is higher than that near the pore wall under other flow conditions.

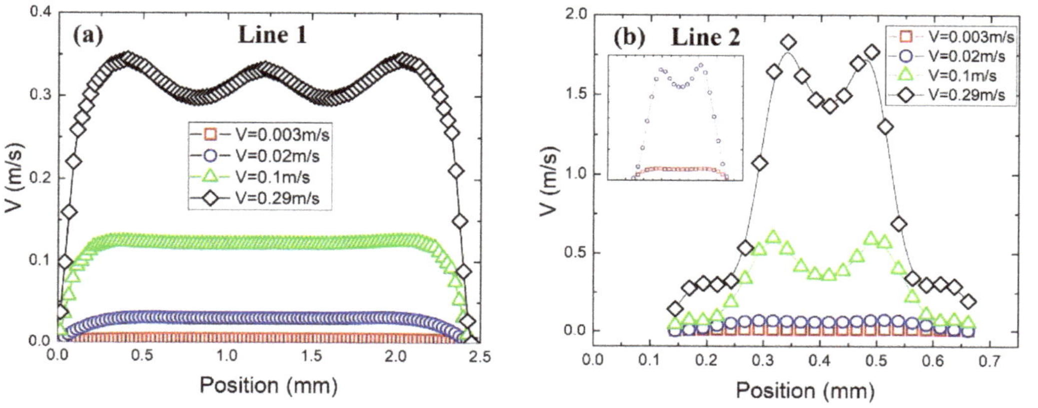

Figure 12. Velocity profiles along a reference line at different flow velocities for the sample with pore size of 550 μm and porosity of 65%. (**a**) Line 1 is located at y = 4.8 × 10^{-3} m and (**b**) line 2 is located at the position marked in Figure 11a.

3.4.2. Effect of Porosity and Pore Size

Porosity and pore size mainly affect the transition velocity (critical Reynolds number) from laminar flow to turbulent flow. Their effect on velocity distribution is insignificant. This has been proved in our previous physical experiments [13,14].

3.5. Heat Transfer Coefficient

The heat transfer performance of metal foam is affected by the structure and flow behavior inside the foam. Figure 13 shows the variation in the heat transfer coefficient with velocity for samples with different pore sizes and porosities. Obviously, the heat transfer coefficient increases exponentially with flow rate. A fast-increasing regime and a slow-increasing regime can be identified over the entire range of flow rates. The fast-increasing regime is mainly in the low flow rate range where heat convection dominates the heat transfer. In the slow-increasing regime where the flow is moving fast, conduction dominates the heat transfer [14]. Figure 13 also shows that low porosity generally has a higher heat transfer coefficient. A red upper limit line and a black lower limit line are plotted in the figure. For a pore size of 250 μm, 550 μm and 1000 μm, the maximum heat transfer difference is 58%, 55% and 45%, respectively.

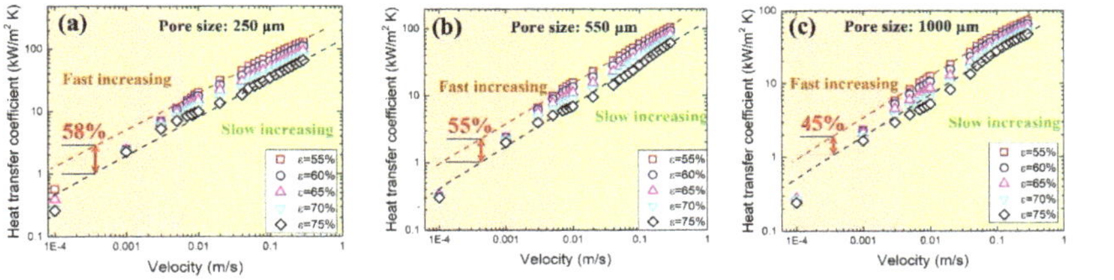

Figure 13. Variation in heat transfer coefficient with flow rate for samples with pore sizes of (**a**) 250 μm, (**b**) 550 μm, (**c**) 1000 μm and different porosities.

Figure 14 plots the heat transfer coefficient as a function of porosity at different velocities. When the velocity is equal to or greater than 0.003 m/s, the heat transfer coefficient decreases with the increase in porosity because the thermal conductivity decreases with the increase in porosity. However, the effect of porosity on heat transfer disappears when the fluid moves at a velocity less than 0.003 m/s, which can be seen from Figures 13 and 14. This is because at low velocity, the heat transferred by conduction through the metal foam is much greater than the heat removed by convection. At extremely low velocity, the intensity of convection is so weak that heat transfer between the metal foam and the fluid, as well as within the fluid, can be considered by conduction. Therefore, the effect of porosity on heat transfer is negligible.

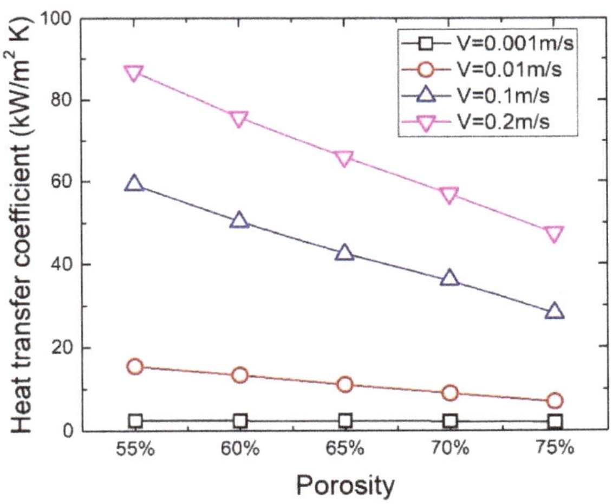

Figure 14. Variation in heat transfer coefficient with porosity for metal foams at different velocities.

Figure 15 shows the effect of pore size on heat transfer coefficients. Foam with a pore size of 250 μm showed the best heat transfer performance, followed by 550 μm and 1000 μm. Similar to the effect of porosity on heat transfer, the effect of pore size is also insignificant at extremely low velocity. In addition, the calculated results are consistent with our previous experimental results, in which the relationship between the heat transfer coefficient and velocity can be divided into three sections, as shown in Figure 15 [14]. As discussed in this section, the changes in the relationship are mainly due to changes in fluid flow behavior.

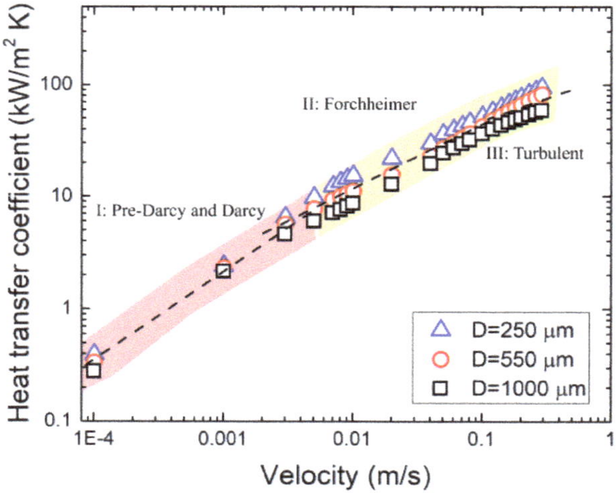

Figure 15. Variation in heat transfer coefficient with pore size for metal foam with different pore sizes.

4. Conclusions

This paper numerically investigated the fluid flow behavior and heat transfer performance of open-cell copper foam by using ANSYS Fluent. A novel geometry model with varying porosity (55–75%) and pore size (250 μm, 550 μm and 1000 μm) was created based on the face-centered cubic arrangement of spheres. The flow velocity ranges from 0.0001 m/s to 0.3 m/s, covering pre-Darcy, Darcy, Forchheimer, and turbulent regimes.

(1) The effect of the number of model layers on the results is insignificant when the number of layers is equal to or greater than two. For the single-layer model, wall effects cannot be ignored, especially at high flow velocity.
(2) The pressure drop increased with decreasing porosity and increasing pore size. The relationship between the pressure drop and flow velocity can be correlated based on the Forchheimer equation. Permeability increased and the form drag coefficient decreased with increasing porosity and both increased with increasing pore size, but the pore size effect on the form drag coefficient is stronger.
(3) The laminar, transitional, and turbulent regimes were identified from the friction factor–velocity diagram. Low porosity and a small pore size results in a large friction factor. Flow in a large-pore-size model tends to become turbulent at a relatively low velocity.
(4) Temperature distribution is largely affected by flow velocity and porosity. Fast-moving fluid and high porosity results in an extremely uneven temperature distribution. Particularly, at high velocity, only a small thickness of metal foam close to the heat source contributes to heat transfer.
(5) Pore-scale velocity is hardly affected by porosity and pore size but is affected by flow rate. At high flow rates, vortices can be found in every pore.
(6) The heat transfer coefficient increased exponentially with flow velocity and increased with decreasing pore size and porosity. However, at extremely low velocity ($\leq$0.001 m/s), the effect of porosity and pore size on heat transfer is negligible.
(7) The numerical results provide useful guidance for the design of metal foam heat exchangers. For example, increasing the thickness of porous media to a certain point no longer increases heat dissipation efficiency when the flow rate is high enough.

Author Contributions: Conceptualization, X.L. and Y.Z. (Yuyuan Zhao); methodology, X.L. and Y.Z. (Yue Zhang); software, X.L.; validation, M.W.; formal analysis, X.L. and Y.Z. (Yuyuan Zhao); investigation, M.W.; writing—original draft preparation, X.L.; writing—review and editing, Y.Z. (Yuyuan Zhao); supervision, Y.Z. (Yuyuan Zhao); project administration, Y.Z. (Yuyuan Zhao); funding acquisition, X.L. All authors have read and agreed to the published version of the manuscript.

Funding: This research was funded by the high-level talent funding of Ningbo University of Technology, grant number 2023KQ018 and 2023KQ003. This research also received funding from the Fenghua Research Institute of Ningbo University of Technology.

Institutional Review Board Statement: Not applicable.

Informed Consent Statement: Not applicable.

Data Availability Statement: The raw data supporting the conclusions of this article will be made available by the authors on request.

Acknowledgments: The authors would like to thank the funding supported by the State Key Laboratory of Powder Metallurgy, Central South University and Ningbo University of Technology.

Conflicts of Interest: Author Yue Zhang is employed by the China Merchants Chongqing Communications Research & Design Institute Co., Ltd. The remaining authors declare that the research was conducted in the absence of any commercial or financial relationships that could be construed as a potential conflict of interest.

References

1. Zhang, Y.; Hao, N.; Lin, X.; Nie, S. Emerging challenges in the thermal management of cellulose nanofibril-based supercapacitors, lithium-ion batteries and solar cells: A review. *Carbohydr. Polym.* **2020**, *234*, 115888. [CrossRef] [PubMed]
2. He, Z.; Yan, Y.; Zhang, Z. Thermal management and temperature uniformity enhancement of electronic devices by micro heat sinks: A review. *Energy* **2021**, *216*, 119223. [CrossRef]
3. Kim, J.; Oh, J.; Lee, H. Review on battery thermal management system for electric vehicles. *Appl. Therm. Eng.* **2019**, *149*, 192–212. [CrossRef]
4. Zhang, X.; Li, Z.; Luo, L.; Fan, Y.; Du, Z. A review on thermal management of lithium-ion batteries for electric vehicles. *Energy* **2022**, *238*, 121652. [CrossRef]
5. Huisseune, H.; De Schampheleire, S.; Ameel, B.; De Paepe, M. Comparison of metal foam heat exchangers to a finned heat exchanger for low Reynolds number applications. *Int. J. Heat Mass Transf.* **2015**, *89*, 1–9. [CrossRef]
6. Samudre, P.; Kailas, S.V. Thermal performance enhancement in open-pore metal foam and foam-fin heat sinks for electronics cooling. *Appl. Therm. Eng.* **2022**, *205*, 117885. [CrossRef]
7. Kuruneru, S.T.W.; Vafai, K.; Sauret, E.; Gu, Y. Application of porous metal foam heat exchangers and the implications of particulate fouling for energy-intensive industries. *Chem. Eng. Sci.* **2020**, *228*, 115968. [CrossRef]
8. Holst, J.R.; Cooper, A.I. *Ultrahigh Surface Area in Porous Solids*; Wiley Online Library: Hoboken, NJ, USA, 2010.
9. Huang, A.; He, Y.; Zhou, Y.; Zhou, Y.; Yang, Y.; Zhang, J.; Luo, L.; Mao, Q.; Hou, D.; Yang, J. A review of recent applications of porous metals and metal oxide in energy storage, sensing and catalysis. *J. Mater. Sci.* **2019**, *54*, 949–973. [CrossRef]
10. Ashby, M.F.; Evans, T.; Fleck, N.A.; Hutchinson, J.; Wadley, H.; Gibson, L. *Metal Foams: A Design Guide*; Elsevier: Amsterdam, The Netherlands, 2000.
11. Banhart, J. Manufacture, characterisation and application of cellular metals and metal foams. *Prog. Mater. Sci.* **2001**, *46*, 559–632. [CrossRef]
12. Zhao, Y.Y.; Fung, T.; Zhang, L.P.; Zhang, F.L. Lost carbonate sintering process for manufacturing metal foams. *Scr. Mater.* **2005**, *52*, 295–298. [CrossRef]
13. Lu, X.; Zhao, Y.; Dennis, D.J. Flow measurements in microporous media using micro-particle image velocimetry. *Phys. Rev. Fluids* **2018**, *3*, 104202. [CrossRef]
14. Lu, X.; Zhao, Y. Effect of flow regime on convective heat transfer in porous copper manufactured by lost carbonate sintering. *Int. J. Heat Fluid Flow* **2019**, *80*, 108482. [CrossRef]
15. Hu, X.; Gong, X. Pore-scale numerical simulation of the thermal performance for phase change material embedded in metal foam with cubic periodic cell structure. *Appl. Therm. Eng.* **2019**, *151*, 231–239. [CrossRef]
16. Panda, J.P.; Warrior, H.V. Numerical studies on drag reduction of an axisymmetric body of revolution with antiturbulence surface. *J. Offshore Mech. Arct. Eng.* **2021**, *143*, 064501. [CrossRef]
17. Panda, J.P.; Kumar, B.; Kumar, M.; Patil, A.K. Influence of twisted tape length on the thermal performance of a heat exchanger tube. *Numer. Heat Transf. Part A Appl.* **2023**, *83*, 650–663. [CrossRef]
18. Lu, T.J.; Stone, H.A.; Ashby, M.F. Heat transfer in open-cell metal foams. *Acta Mater.* **1998**, *46*, 3619–3635. [CrossRef]
19. Boomsma, K.; Poulikakos, D.; Ventikos, Y. Simulations of flow through open cell metal foams using an idealized periodic cell structure. *Int. J. Heat Fluid Flow* **2003**, *24*, 825–834. [CrossRef]

20. Kopanidis, A.; Theodorakakos, A.; Gavaises, E.; Bouris, D. 3D numerical simulation of flow and conjugate heat transfer through a pore scale model of high porosity open cell metal foam. *Int. J. Heat Mass Transf.* **2010**, *53*, 2539–2550. [CrossRef]
21. Hu, C.; Sun, M.; Xie, Z.; Yang, L.; Song, Y.; Tang, D.; Zhao, J. Numerical simulation on the forced convection heat transfer of porous medium for turbine engine heat exchanger applications. *Appl. Therm. Eng.* **2020**, *180*, 115845. [CrossRef]
22. Yang, K.; Liu, K.; Wang, J. Pore-scale numerical simulation of convection heat transfer in high porosity open-cell metal foam under rotating conditions. *Appl. Therm. Eng.* **2021**, *195*, 117168. [CrossRef]
23. Gauna, E.A.; Zhao, Y. Numerical Simulation of Heat Transfer in Porous Metals for Cooling Applications. *Metall. Mater. Trans. B* **2017**, *48*, 1925–1932. [CrossRef]
24. Zafari, M.; Panjepour, M.; Meratian, M.; Emami, M.D. CFD simulation of forced convective heat transfer by tetrakaidecahedron model in metal foams. *J. Porous Media* **2016**, *19*. [CrossRef]
25. Zhao, Y. Stochastic modelling of removability of NaCl in sintering and dissolution process to produce Al foams. *J. Porous Mater.* **2003**, *10*, 105–111. [CrossRef]
26. ANSYS Fluent, Relese 16.0. ANSYS, Inc. Available online: http://www.ansys.com/products/fluids/ansys-fluent (accessed on 1 May 2019).
27. Zhao, C.; Tassou, S.; Lu, T. Analytical considerations of thermal radiation in cellular metal foams with open cells. *Int. J. Heat Mass Transf.* **2008**, *51*, 929–940. [CrossRef]
28. Baloyo, J.M.; Zhao, Y. Permeability, Form Drag Coefficient and Heat Transfer Coefficient of Porous Copper. *Heat Transf. Eng.* **2021**, *42*, 670–682. [CrossRef]
29. Darcy, H. *LES FONTAINES PUBLIQUES de la Ville de Dijon: Exposition et Application des Principes à Suivre et des Formules à Employer Dans les Questions de Distribution D'eau: Ouvrage Terminé par un Appendice Relatif aux Fournitures D'eau de Plusieurs Villes, au Filtrage des Eaux et à la Fabrication des Tuyaux de Fonte, de Plomb, de Tôle et de Bitume*; V. Dalmont: Paris, France, 1856; Volume 2.
30. Diao, K.; Zhang, L.; Zhao, Y. Measurement of tortuosity of porous Cu using a diffusion diaphragm cell. *Measurement* **2017**, *110*, 335–338. [CrossRef]
31. Solórzano, E.; Reglero, J.A.; Rodríguez-Pérez, M.A.; Lehmhus, D.; Wichmann, M.; de Saja, J.A. An experimental study on the thermal conductivity of aluminium foams by using the transient plane source method. *Int. J. Heat Mass Transf.* **2008**, *51*, 6259–6267. [CrossRef]
32. Kumar, P.; Topin, F. Different arrangements of simplified models to predict effective thermal conductivity of open-cell foams. *Heat Mass Transf.* **2017**, *53*, 2473–2486. [CrossRef]
33. Gopinathan, A.; Jerz, J.; Kováčik, J.; Dvorák, T. Investigation of the relationship between morphology and thermal conductivity of powder metallurgically prepared aluminium foams. *Materials* **2021**, *14*, 3623. [CrossRef]

materials

MDPI

Article

Energy Absorption Characteristics of Composite Material with Fiber–Foam Metal Sandwich Structure Subjected to Gas Explosion

Baoyong Zhang, Jin Tao *, Jiarui Cui, Yiyu Zhang, Yajun Wang, Yingxin Zhang, Yonghui Han and Man Sun

Department of Safety Engineering, Heilongjiang University of Science and Technology, Harbin 150022, China; byzhang1982@163.com (B.Z.); 19835127449@163.com (J.C.); m18846190121@163.com (Y.Z.); 2006800307@usth.edu.cn (Y.W.); zhangyingxin01@163.com (Y.Z.); hanyonghui1@usth.edu.cn (Y.H.); 18745000523@163.com (M.S.)
* Correspondence: taojin@usth.edu.cn; Tel.: +1-339-451-5557

Abstract: Based on the previous research on the energy absorption of foam metal materials with different structures, a composite blast-resistant energy-absorbing material with a flexible core layer was designed. The material is composed of three different fiber materials (carbon fiber, aramid fiber, and glass fiber) as the core layer and foamed iron–nickel metal as the front and rear panels. The energy absorption characteristics were tested using a self-built gas explosion tube network experimental platform, and the energy absorption effects of different combinations of blast-resistant materials were analyzed. The purpose of this paper is to evaluate the performance of blast-resistant materials designed with flexible fiber core layers. The experimental results show that the composite structure blast-resistant material with a flexible core layer has higher energy absorption performance. The work performed in this paper shows that the use of flexible core layer materials has great research potential and engineering research value for improving energy absorption performance, reducing the mass of blast-resistant materials, and reducing production costs. It also provides thoughts for the research of biomimetic energy-absorbing materials.

Keywords: gas explosion; core layer; foamed metal; explosion-proof capability; composite

Citation: Zhang, B.; Tao, J.; Cui, J.; Zhang, Y.; Wang, Y.; Zhang, Y.; Han, Y.; Sun, M. Energy Absorption Characteristics of Composite Material with Fiber–Foam Metal Sandwich Structure Subjected to Gas Explosion. *Materials* **2024**, *17*, 1596. https://doi.org/10.3390/ma17071596

Academic Editor: Eddie Koenders

Received: 5 February 2024
Revised: 9 March 2024
Accepted: 26 March 2024
Published: 31 March 2024

1. Introduction

Gas and other dangerous gas explosions are common hazards in industrial and mining enterprises. The key to rescue and relief is to reduce casualties after the explosion. Currently, explosion suppression methods can be broadly divided into gas explosion suppression, liquid explosion suppression, and solid explosion suppression [1].

Gas explosion suppression mainly employs experimental and numerical simulation methods from a macroscopic point of view to study the role of a single or complex gas explosion suppression system. The research objects are mainly inert gases, N_2, CO_2, and other gases. Liquid explosion suppression generally involves the addition of aqueous media by changing the spray molecular volume, additives, and charge of fine water mist on the gas explosion pressure and flame propagation velocity inhibition effect. In their research on solid explosion suppression, many scholars have found that porous materials have unique physical and mechanical properties and excellent energy absorption performance. They have carried out extensive research, and these materials have been widely used in the field of engineering protection.

Foam metal is one of the research hotspots of scholars due to its advantages of low density, large specific surface area, and high thermal conductivity. Zhuang et al. [2] conducted an experimental study on the suppression effect of different porous materials on the explosion of combustible gases. The results showed that the thickness and pore size changes in the composite porous material have a great influence on the explosion pressure and explosion intensity. Varun et al. [3] conducted a numerical simulation of three samples with different porosities (30%, 60%, and 80%) uniformly compressed in the uniaxial (Z-axis) direction to study the influence of porosity on the mechanical properties of open-pore Voronoi foam. Wang Yajun et al. [4] found through their personally designed experimental device that when the volume density of foam metal is higher, its explosion-proof performance is better, but adding a certain amount of coal dust to the explosion device will reduce the explosion-proof performance of foam metal. Wei Chunrong et al. [5] used a self-designed and processed square explosion experimental pipeline with a cross section of 30 cm × 30 cm to compare the explosion-proof effects of metal wire mesh, foam ceramics, and foam iron–nickel metal with different parameters. Yu Minggao et al. [6] studied the influence law of the synergistic effect of ultra-fine water mist and foam metal on explosion overpressure and found that changing the parameters, such as material porosity, can improve the explosion-proof effect of the experiment. Zhang Baoyong et al. [7] conducted an experimental study on the energy absorption performance of explosion-proof materials with a sawtooth structure and analyzed the influence of surface structure on energy absorption performance.

In addition to researching foam metal as a barrier material, scholars have also increasingly focused their attention on multilayer sandwich structures [8–17]. Tarlochan, F. [18] discussed the use of sandwich structures in energy absorption applications and found that sandwich structures are a good choice for energy absorbers. It is suggested that the way forward is to design sandwich structures by using a combination of "artificial intelligence/data mining and topology optimisation." Many scholars have used sandwich structures designed with foamed metal in order to obtain higher energy absorption performance [3,19–31]. Zunjarrao, K. [32] reviewed current research on innovative sandwich structures, including integral woven corrugated cores, honeycomb cores, foam cores, and 3D printed core structures, and highlighted their versatility. Mao [33] used theoretical and numerical methods to study the attenuation of shock waves generated by gas explosions by an energy absorption device composed of aluminum foam and steel plates. The results showed that the multilayer composite structure has a good ability to reduce explosion load and attenuate air shock wave overpressure. The foam aluminum layer has the ability to attenuate explosion pressure before compression, and the attenuation ability decreases after the foam aluminum is completely compressed. Zhang et al. [34] studied and numerically analyzed the "effective" compressive strength and the dynamic response of corrugated sandwich panels with unfilled and foam-filled sinusoidal corrugated cores. The dynamic response of fully supported sandwich panels with unfilled and foam-filled sinusoidal corrugated cores under impact loading was analyzed using the finite element method. Chen et al. [35] carried out a study of the explosion protection properties of a composite structural barrier material with polymer interlayers using the LS-DYNA software and calibrated numerical models to simulate the explosion resistance and energy absorption capacity of the composite material under long-distance explosive loading conditions. Zhou et al. [36] predicted the compressive strength and dynamic response of a corrugated sandwich panel consisting of a panel and a metal foam core in close proximity to explosions. An energy-based method was proposed to predict the depth and scope of deformation in the outer panel, providing insights for the design and sizing of the core layer material. Langdon et al. [37] classified the energy absorption properties and applications of fiber-reinforced polymer composites and studied the multilayered (including the sandwich structure) and mixed composite metal structure materials. The numerical predictions were compared and analyzed with experimental data. In order to improve the protective performance of aluminum foam sandwich composites against blast shock wave and fragmentation penetration, Zhou et al. [38] studied the damage modes and mechanisms of the structure using the "blast + fragmentation" intrusion experiments. The effects of core material combination and blast distance on protective performance were also discussed and verified by the LS-DYNA numerical simulation. Santosa et al. [39] investigated the blast resistance and barrier performance of metal foam sandwich panels with different thicknesses, materials, and bulk densities by changing the impact distance between the blast load and the foam sandwich panels. Guo et al. [40] designed a metal foam-filled sandwich cylinder (MFSC) and found that adjusting the thickness of the foam based on the impact distance of the foam sandwich panels had no significant effect on the barrier performance by means of experiments and finite element calculations. In contrast, a finite element calculation found that adjusting the ratio of foam thickness to tube wall thickness can improve the load-carrying capacity and energy-absorption capacity of the energy-absorbing structure.

Inspired by the ancient Chinese philosophical idea of "combining hardness and softness" and the "sandwich" core structure, and based on the previous experimental study on the energy absorption characteristics of different explosion-proof surface materials, a composite explosion-proof and energy-absorbing material with foam metal as the upper and lower panels and fiber material as the core layer was designed. The explosion overpressure, flame propagation speed, flame temperature, and other characteristic parameters of the front and rear ends of the composite material after being subjected to a methane–air mixture gas explosion were collected and analyzed. The research results are expected to provide an experimental and analytical basis for the application of fiber–foam metal sandwich structure composite materials in the fields of gas explosion isolation and energy absorption.

2. Experimental

2.1. Materials

The experimental materials included foamed metal ferronickel, carbon fiber, aramid fiber, and glass fiber. The foam iron–nickel metal was produced by Jilin Zhuoer Metal Material Preparation Company (Jilin City, China). Pore density was 30, bulk density was 0.4–0.5 g/cm^3, tensile strength was not less than 50 MPa, and compressive strength was not less than 250 kPa.

The fiber materials were all made of tough fiber fabrics from Zhongfu Fiber Textile Preparation Co., Ltd. (Suqian City, China) The fabric density was 1000 D, and the single-layer thickness was 1 mm. Five layers of each of the three materials, carbon fiber, aramid fiber, and glass fiber, were selected, and the core thickness was 5 mm.

2.2. Equipment

As shown in Figure 1, the experimental system consists of six main parts in turn: the gas distribution system, ignition system, explosion chamber, explosion diffusion line, material gripper, and data acquisition system [41]. The end of the explosion chamber and the diffusion line are equipped with a PTFE film with a thickness of 0.3 mm and a breaking pressure of 90 kPa, and the edge of the gripper is sealed by a rubber ring, which is used to ensure the hermeticity inside the chamber. The main equipment of the gas distribution system is a vacuum pump. Under normal temperature and pressure conditions, the vacuum pump extracts part of the air from the explosion chamber to form a relative negative pressure state in the chamber. The required combustible gas is filled into the explosion chamber in this state. According to the experimental requirements, a certain volume of pure methane gas is charged to make it a mixed gas with an explosion concentration of 9.5%. An electric spark generator with an ignition voltage of 220 V and an ignition energy of 440 J is installed at the front end of the explosion chamber. The ignition electrode is responsible for remote charging and ignition by the terminal ignition system. The back end of the explosion chamber is sealed by a circular steel plate with a diameter of 300 mm and a thickness of 40 mm, and there is a hole of 118 mm in the center of the circular steel plate to connect with the diffusion line at the back end. The inner diameter of the diffusion pipe is 118 mm. It is spliced together from three hollow steel pipes with a single pipe length of 2200 mm. The total length of the explosion diffusion pipe is 6600 mm.

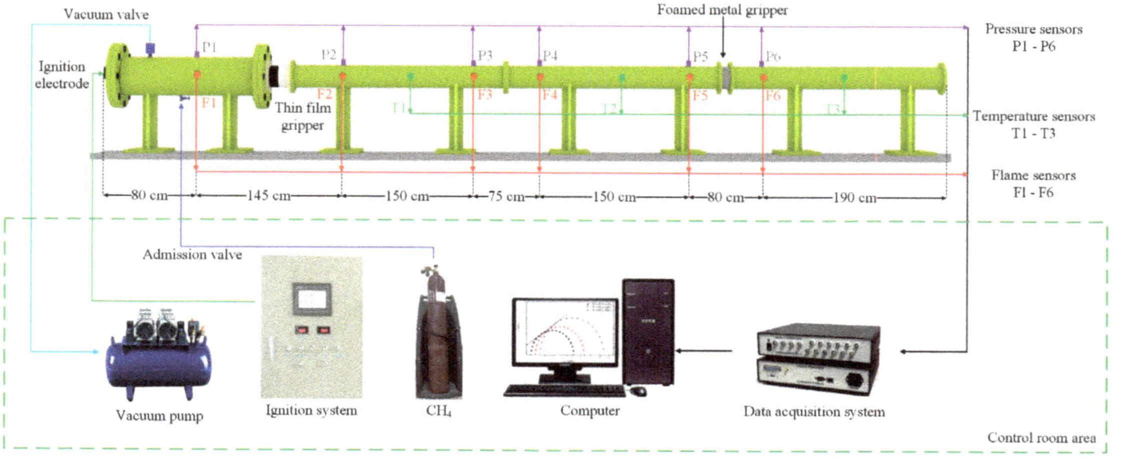

Figure 1. Experimental system.

Flame sensors and pressure sensors, numbered 1–6, and temperature sensors, numbered 1–3, are installed at different locations in the experimental pipeline. The pressure sensor, flame sensor, and temperature sensor are all provided by Chengdu Tester Company (Chengdu, China), of which the pressure sensor model is CT100T, with a sensor range of 0~2 MPa; the flame sensor model is CKG100, with a response spectrum of 450 nm~980 nm; the temperature sensor model is C2 fast-response thermocouple, with a sensor range of 0~2500 °C; and the response time of all three sensors is less than 100 μs. Foam metal is installed in the material gripper, which is located between the No. 5 and No. 6 flame sensors and pressure sensors and is 0.15 m away from the No. 5 sensors. The data acquisition system used is the TST6300 data acquisition system from Chengdu Tester Company. The data acquisition objects are flame propagation speed, explosion overpressure, and flame temperature. The data sampling frequency is 100 kHz. The acquisition method is internal triggering. The recording time starts from the trigger ignition, and the entire recording process is less than 2 s.

2.3. Experimental Steps and Program

This experiment uses the explosion tube network device shown in Figure 1 to test the explosion-resistant performance of materials with different blast-facing structures by means of a gas explosion. The specific experimental process is as follows: First, clean the explosion chamber and explosion pipe network, seal the connection between the explosion chamber and the explosion-proof film holder with a 0.3 mm thick film, and seal the edge of the holder with a sealing ring around the explosion-proof film holder. Subsequently, the explosion chamber with a length of 1600 mm and a diameter of 300 mm is filled with a certain volume of methane gas, so that the interior of the chamber is filled with a methane–air gas mixture with a volume fraction of 9.50% for the explosion experiment. During the whole experiment, the data acquisition system collects data from different sensors at each measuring point for comparison, so as to observe the changing patterns of the impeded change in flame temperature propagation, the change in explosion overpressure before and after the material, and the flame propagation in the pipeline as an indicator to judge the explosion-resistant performance of the material.

The experimental conditions are shown in Table 1. The front and rear panels of the explosion-proof material are both foam iron–nickel metal. The front panel is surface-modified according to the experimental design requirements. Before the experiment, the wire cutting method is used to prepare the explosion-facing surface into a sawtooth wave with a thickness of 5 mm and an angle of 30°. The fiber core layer is divided into three working conditions: carbon fiber, aramid fiber, and glass fiber. In order to improve the data comparison effect of each working condition, experiment 1 is set as the control experiment. The thickness of the front panel is 15 mm, and no fiber material is added in the middle. The material parameters of other experiments are based on Table 1. In addition, in order to test the stability of the experimental platform, an empty pipe experiment without explosion-proof material is added.

Table 1. Design parameters of external explosive surface tests.

Number of Test	Fiber Type	Foamed Metal	Body Density (g/cm^3)	Serrated Thickness (mm)	Material Thickness (mm)
1	—	foamed ferronickel	0.5	5	15 + 0 + 10
2	carbon	foamed ferronickel	0.5	5	10 + 5 + 10
3	aramid	foamed ferronickel	0.5	5	10 + 5 + 10
4	glass	foamed ferronickel	0.5	5	10 + 5 + 10

The experiment is mainly conducted using the three aspects of the research data—the explosion overpressure, flame propagation speed, and flame temperature—as well as a comparative evaluation of different experimental materials on the explosion overpressure and other parameters of inhibition. The explosion overpressure, flame propagation, speed, and flame temperature-related formulas [41] are shown in (1)–(5):

$$V = \mathrm{d}p/\mathrm{d}t = (p_{\max} - p_{\mathrm{i}})/\Delta t = \Delta p/\Delta t \tag{1}$$

where V is the rate of decline in overpressure, in MPa/s; $P_{\max}$ is the front end of the test material maximum explosion overpressure, in MPa; P_{i} is the back end of the test material maximum explosion overpressure, in MPa; Δp is the test material before and after the two ends of the explosion pressure difference, in MPa; and Δt is the sensor detection signal time difference, in seconds.

$$\zeta = (p_{\max} - p_{\mathrm{i}})/p_{\max} \tag{2}$$

where $P_{\max}$ is the experimental material in the front channel of the maximum overpressure, in MPa; P_{i} is the explosion conditions of the experimental material at the back end of the maximum explosion overpressure, in MPa; and ζ is the overpressure attenuation rate, that is, the material of the maximum overpressure reduction control ability.

The blocking effect of the experimental material on the flame propagation velocity can be compared with the flame propagation velocity decay rate as follows:

$$\mu = \Delta v/v_{\max} \tag{3}$$

where μ is the foam metal before and after the flame propagation velocity attenuation rate, in m/s; Δv is the foam metal before and after the difference in velocity propagation, in m/s; and $v_{\max}$ is the foam metal during the explosion before the end of the maximum value of velocity, in m/s.

The damping effect of the experimental materials on the flame temperature can be compared with the flame temperature decay rate as follows:

$$\eta = (T_{\mathrm{f,max}} - T_{\mathrm{i,max}})/T_{\mathrm{f,max}} \tag{4}$$

where $T_{\max}$ is the maximum temperature at the front end of the experimental material, in °C; T_{i} is the maximum temperature at the back end of the experimental material, in °C; and η is the flame temperature attenuation rate, i.e., the material's abatement control ability for the maximum temperature.

The overall protection effect of the experimental materials can be compared with the explosion quenching parameters as follows:

$$\theta = TP \tag{5}$$

where θ is the quenching parameter, in MPa·°C; T is the flame temperature, in °C; and P is the explosion overpressure, in MPa.

3. Results and Discussion

3.1. Comparative Study of the Blast Overpressure Barrier Effect

Figure 2 shows the explosion overpressure–distance data before and after different experiments. The maximum explosion overpressure of experiments 1 to 4 is reduced to 0.105 MPa, 0.067 MPa, 0.069 MPa, and 0.034 MPa, respectively, after passing through the explosion-proof material. The sensor value is significantly reduced after the explosion-proof material, and it can be clearly seen from the experimental data that experiments 2 to 4 have a better pressure drop effect under the same gas explosion overpressure propagation conditions [41] and are better than the single foam metal structure explosion-proof material.

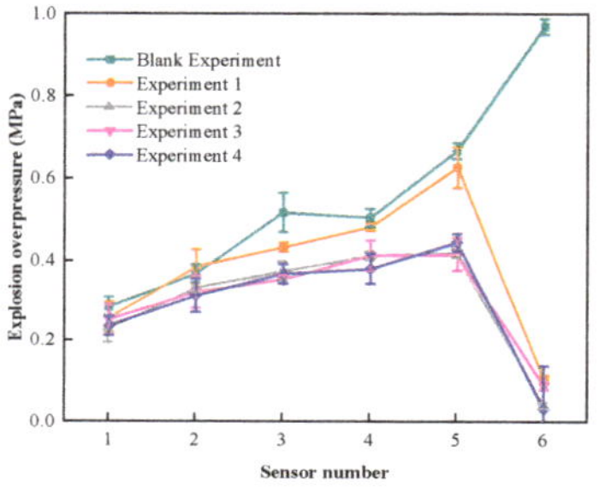

Figure 2. Explosive overpressure–distance data.

At the same time, the overpressure attenuation results of the three fiber material core explosion-proof materials have large differences: the overpressure attenuation rate of the carbon fiber–foam metal explosion-proof material to gas explosion overpressure is 85.27%, and the explosion overpressure before and after the material is reduced by 0.388 MPa. The overpressure attenuation rate of aramid fiber–foam metal explosion-proof material to gas explosion overpressure is 84.24%, and the explosion overpressure before and after the material is reduced by 0.369 MPa. The overpressure attenuation rate of glass fiber–foam metal explosion-proof material is 92.01%, and the explosion overpressure before and after the material is reduced by 0.392 MPa. From the perspective of the overpressure attenuation rate of different core layer explosion-proof materials, glass fiber has the best effect, followed by aramid fiber, and carbon fiber has the lowest effect.

The evolution of explosion overpressure before and after different composite materials is shown in Figure 3. According to the data analysis in the figure, the explosion overpressure drop rates of experiments 1 to 4 are 11.82 MPa/s, 3.62 MPa/s, 4.11 MPa/s, and 2.78 MPa/s, respectively. There are two reasons why the explosion overpressure drop rates of experiments 2 to 4 are smaller than for experiment 1: (1) the maximum explosion overpressure detected by sensor P5 in experiments 2 to 4 is smaller than that of experiment 1; (2) due to the addition of tough fibers, the material hinders the propagation of the explosion shock wave, prolongs the time for the explosion shock wave to pass through the entire material, and thus plays a role in weakening the penetration ability of the explosion shock wave [41].

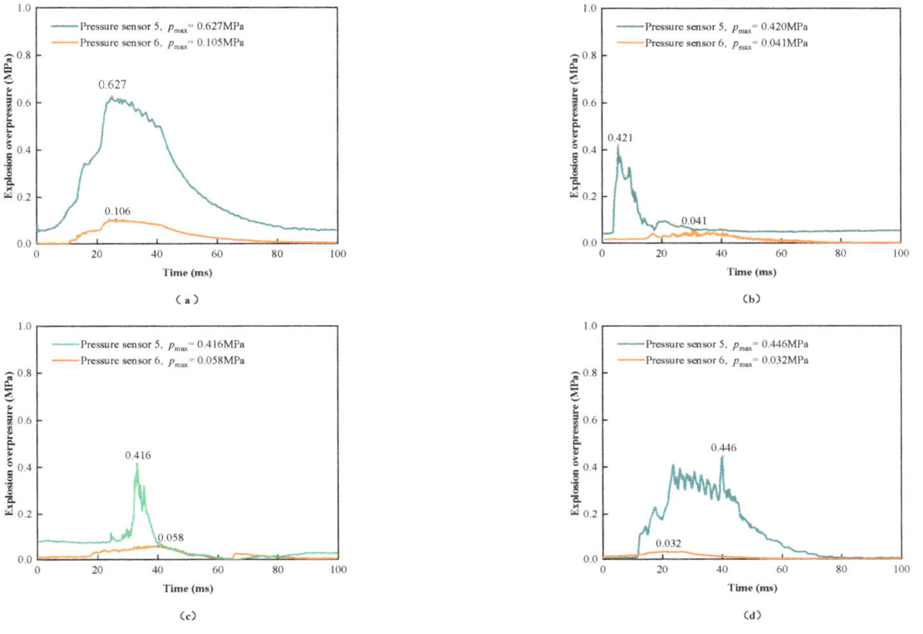

Figure 3. Different composite material overpressure change rule with time. (**a**) Without a core layer. (**b**) Carbon fiber core layer. (**c**) Aramid fiber core layer. (**d**) Glass fiber core layer.

Through analysis, it was concluded that when the explosion shock wave propagates inside the foam metal, due to the porous structure characteristics of the foam metal itself, the shock wave overpressure can be divided into several small parts when it passes through the foam metal material, and the pressure can be quickly transmitted and absorbed, reducing the propagation speed of the shock wave and achieving the effect of blocking the propagation of the explosion overpressure [41]. This experiment increases the explosion-proof performance of foam metal under explosion overpressure by adding fiber materials inside the foam metal so as to further weaken the explosion shock. When the explosion overpressure hits the fiber material, due to the fact that the fiber material is a flexible material with better deformation ability, the fiber material will deform and play a buffering role in the process of being subjected to the explosion shock, further protecting the rear material from the stress it bears, improving the overall energy absorption characteristics of the material, and further reducing the explosion shock at the rear of the material. Under the premise of the same foam metal as rigid support and the same surface porosity, glass fiber has higher fiber density, better anti-deformation ability, and absorbs more shock wave overpressure released by gas explosions.

3.2. Comparative Study of the Flame Propagation Blocking Effect

The relationship between flame propagation speed and propagation distance is shown in Figure 4. In experiments 1 to 4, the speeds detected by the sensors at the rear of the explosion-proof material are 65.28 m/s, 57.32 m/s, 52.18 m/s, and 33.48 m/s, respectively. After calculation, the flame propagation speed decreases by 216.27 m/s, 164.17 m/s, 151.09 m/s, and 160.30 m/s, respectively. The results show that adding a fiber material core layer to the foam iron–nickel panel can absorb and suppress the further propagation of the flame, thereby achieving the purpose of improving the explosion-proof effect. The reason for this is that when the flame passes through the fibers inside the foam metal, a "channel" that is smaller and denser than the original foam metal is formed. When the flame flows in the denser "channel," the oxygen supply of the flame is further blocked, achieving the purpose of more easily isolating the flame propagation [41,42]. Similar to the experimental results of explosion overpressure suppression, glass fiber core layer material has the best flame propagation suppression effect.

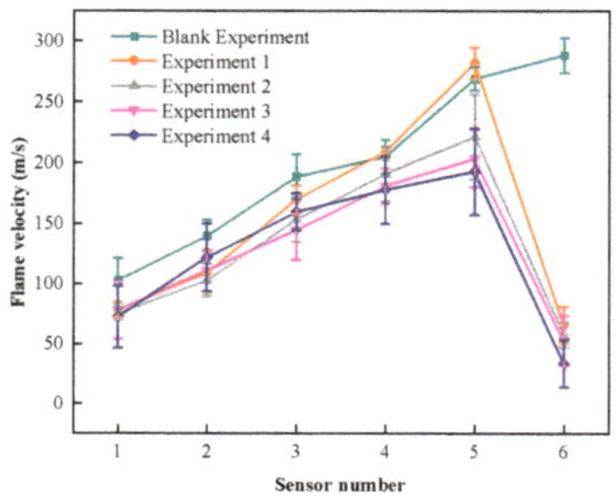

Figure 4. Flame speed–distance data.

Figure 5 shows a graph of the flame propagation velocity suppression effect data. The flame propagation velocity detected at the rear end of the flame retardant material was reduced for all conditions, with the attenuation rate ranging from 74.12% to 89.60%.

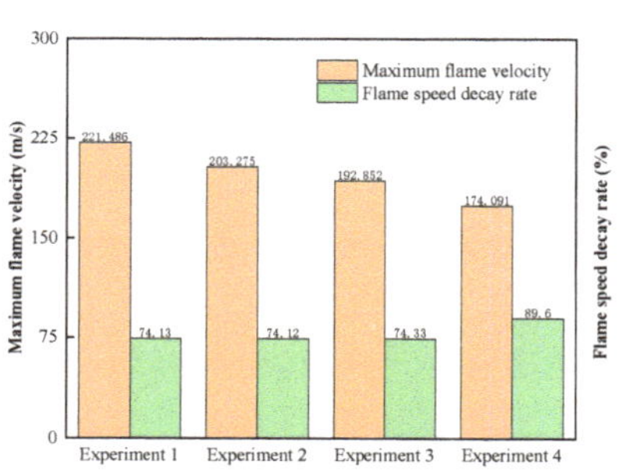

Figure 5. Flame velocity suppression effect data.

The flame propagation speed attenuation rate of the explosion-proof material before and after experiment 4 is 89.60%, the maximum flame propagation speed is 174.091 m/s, and it has the best blocking effect on the explosion flame propagation. The flame propagation speed attenuation rates of experiments 1 to 3 are similar, at 74.13%, 74.12%, and 74.33%, respectively. The maximum flame propagation speeds are 221.486 m/s, 203.275 m/s, and 192.852 m/s, respectively. The maximum flame propagation speeds of experiments 2 to 4 are all lower than for experiment 1 to different degrees, which indicates that the protection effects of experiments 2 to 4 are all better than experiment 1 under the impact of flame.

Figure 6 shows the flame propagation effect diagram at the end of the explosion pipeline. It can be seen from the pictures of experiment 4 that only part of the smoke and gas are diffused from the end of the pipeline, and there is no situation where the material or flame rushes out of the end of the explosion pipeline. The experimental effect is the best. In experiment 3, only a large amount of smoke was sprayed out under the premise of ensuring the integrity of the experimental material. The experimental effect is better. In experiment 2, a small amount of carbon fiber material was sprayed out of the pipeline along with smoke, and the integrity of the core material was not guaranteed. The experimental effect is the worst. By comparing the explosion overpressure and flame propagation results of each working condition and combining the analysis of the actual effect at the end of the experimental pipeline, it is once again demonstrated that adding tough fiber materials to foam metal can effectively improve the overall explosion-proof performance of the material compared to single foam metal experiments.

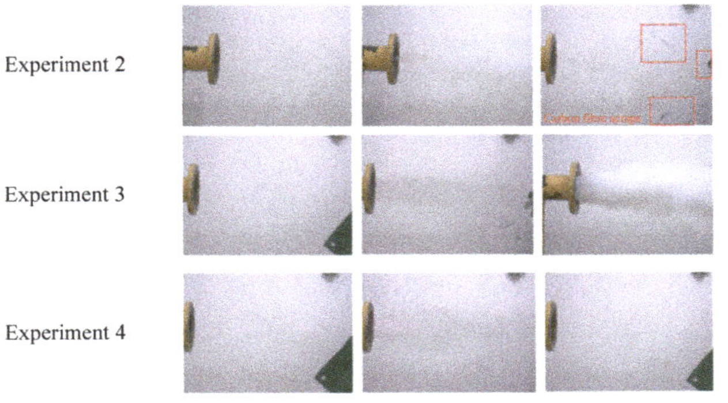

Figure 6. Fire propagation effect diagram at the end of the explosion pipe network.

3.3. Comparative Study of the Flame Temperature Barrier Effect

Figure 7 shows the flame temperature–distance data before and after different explosion-proof materials. As can be seen from the figure, the overall trend of flame temperature propagation is similar to that of explosion overpressure and flame propagation speed. After adding flexible fiber materials to the foam metal material, the explosion-proof effect is better than that of a single foam metal material. Glass fiber has the best effect on the explosion-proof material with a fiber core layer, followed by aramid fiber, and carbon fiber has the worst effect on the explosion-proof material.

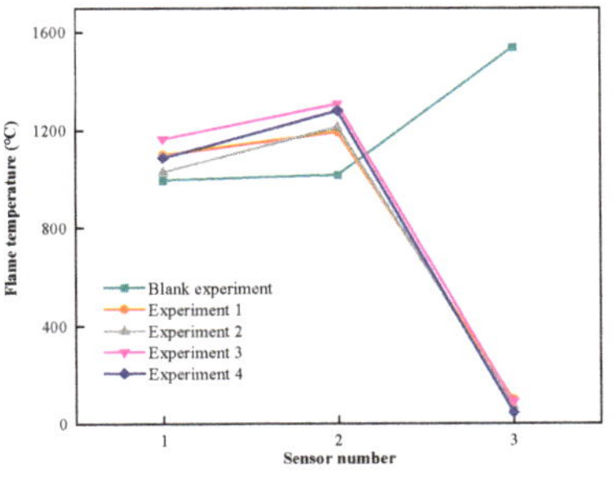

Figure 7. Flame temperature–distance data.

The temperature values detected at the rear of the explosion-proof materials in experiments 1 to 4 are 96.112 °C, 69.578 °C, 85.770 °C, and 44.875 °C, respectively. The flame temperature drop amplitudes are 1094.928 °C, 1141.572 °C, 1219.490 °C, and 1232.695 °C, respectively. The temperature attenuation rates are 91.93%, 94.26%, 93.43%, and 96.49%, respectively. It can be seen from the experimental results of flame temperature attenuation and temperature change that the explosion-proof materials composed of three different fiber core layer materials are still slightly better than the single foam metal material in terms of flame temperature suppression, and the experimental results of the glass fiber core layer are still the best.

Figure 8 shows the comparison of the extinction parameters of different explosion-proof surfaces. The extinction parameter at the rear of the explosion-proof material in experiment 4 is 1.53 MPa·°C, and the explosion-proof effect is the best. The extinction parameters at the rear of the explosion-proof materials in experiments 1 to 3 are 10.09 MPa·°C, 4.66 MPa·°C, and 7.63 MPa·°C, respectively. The values of the extinction parameters before and after the materials are 755.75 MPa·°C, 546.41 MPa·°C, 564.07 MPa·°C, and 542.72 MPa·°C, respectively. When the extinction parameter is lower than 390 MPa·°C, people and equipment are within the relative safety limit, and the lower the extinction parameter, the better the protection effect. Data analysis shows that experiment 1 has a larger extinction parameter drop because the explosion-proof material has a larger initial explosion overpressure value at the front end, which leads to an increase in the extinction parameter. The extinction parameter values at the rear of the materials in experiments 2 to 4 are lower than in experiment 1, indicating that this core layer design can more effectively reduce the damage to the protected target caused by explosion shock, flame, and temperature.

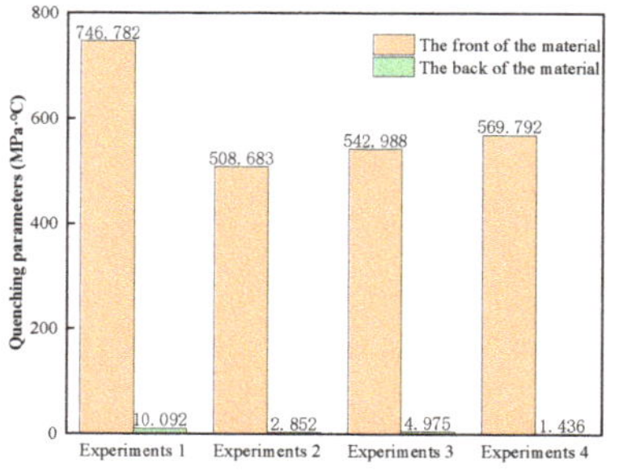

Figure 8. Extinction parameter data at the front and rear of the material.

3.4. Analysis of the Effect of Different Composite Materials to Prevent the Explosion

Experimental studies have found that adding different types of fiber materials to foam metals can effectively improve the overall explosion-proof performance of the experimental materials. Figure 9 shows a schematic diagram of a fiber–foam metal core structure composite material. When the explosion shock wave and flame impact the explosion-proof surface of the foam metal, the explosion and flame impact will penetrate the interior along the sawtooth of the explosion-proof surface. The explosion shock wave and flame are highly concentrated in this area and cause deformation, energy reflection, and scattering at the wall surface when entering the foam metal, thereby achieving the effect of making the material explosion-proof. Adding different types of fiber materials as the core layer in the material can form a "channel" that is smaller and denser than the original foam metal fine pores, which increases the propagation path, further inhibits the occurrence of chemical reactions, and buffers the foam metal while the deformation makes the experimental material absorb more energy holistically, thereby achieving the purpose of improving the explosion-proof effect.

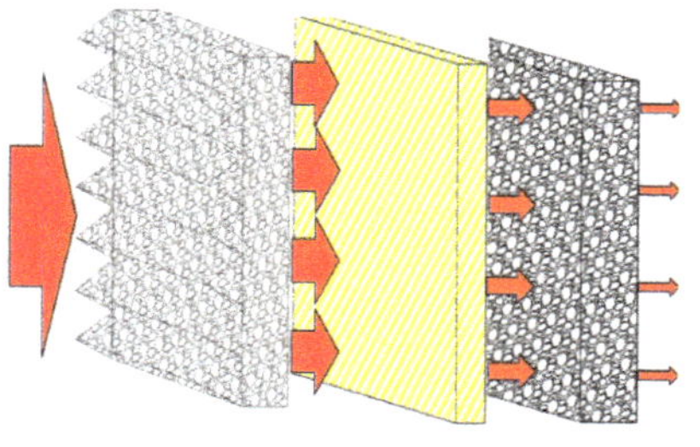

Figure 9. Schematic diagram of the fiber–foam metal core structure composites.

In the experiment, the overall explosion-proof effect of adding glass fiber to foam metal is better than that of other fiber material experiments. The preliminary analysis shows that the fiber density of glass fiber material is higher, so the overall explosion-proof effect of glass fiber is better.

In order to observe more intuitively the state of the foamed FeNi metal before and after the passage of the overpressure shockwave and flame, a scanning electron microscope (SEM) was used to take pictures of the foamed metal before and after the experiments (Figure 10). The changes in the foamed metal material in the microscopic view can clearly be distinguished, which will help with future research on energy absorption in foamed metal. A comparison of the images shows that the surface structure of the foamed metal received damage and became rougher after the shockwave overpressure and the passage of the flame, along with traces of cracking, melting, ablation, and material detachment, which changed the pore structure of the foamed metal. When the surface of the foamed metal was observed at magnification, the surface changes were evident after the passage of the flame, presumably as a result of oxidation or other chemical reactions.

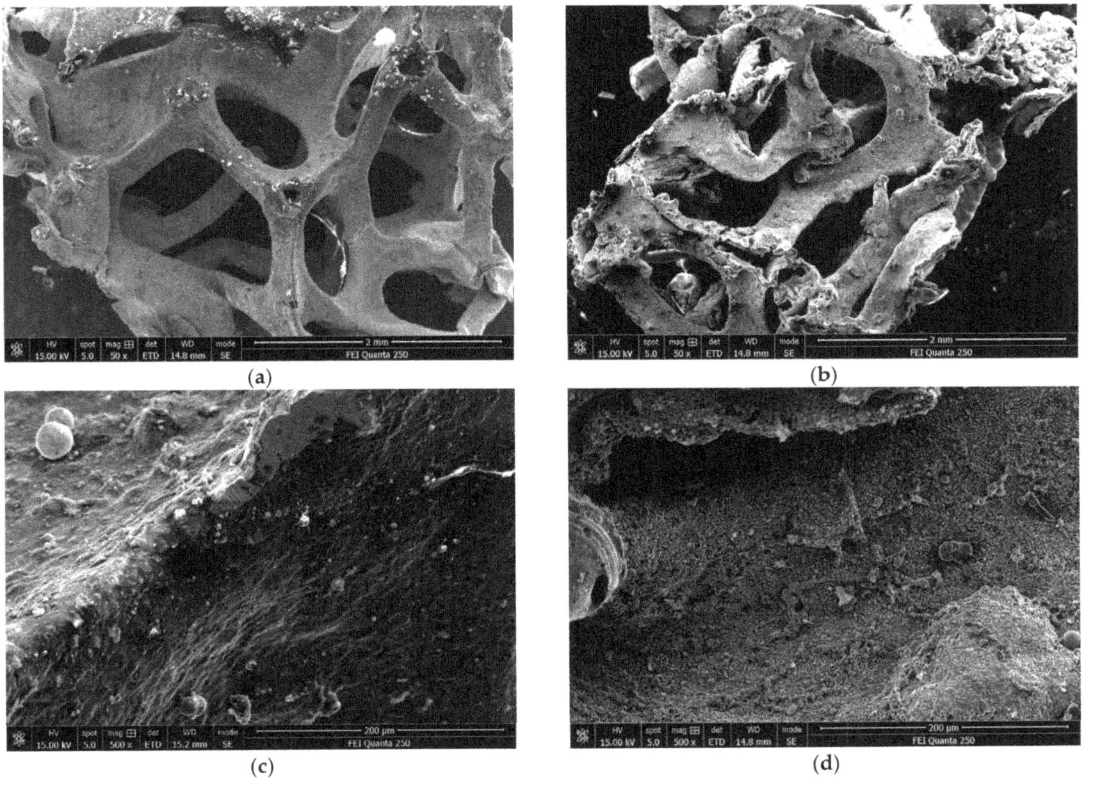

Figure 10. *Cont.*

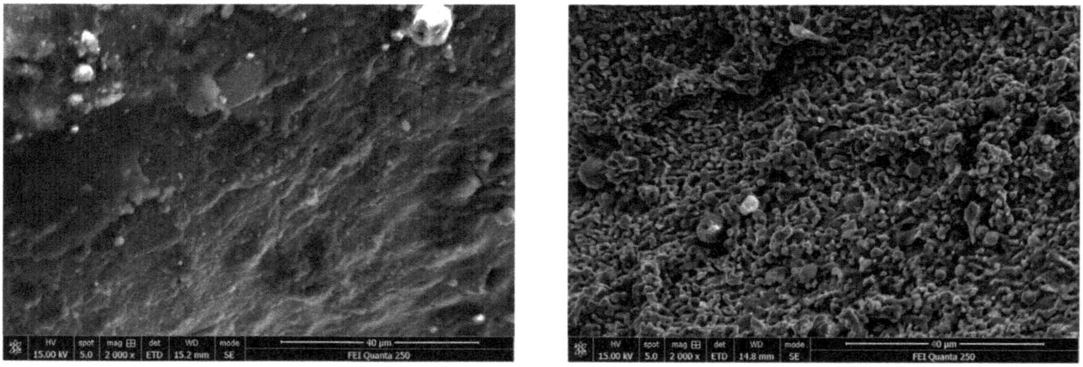

(e) (f)

Figure 10. Scanning electron microscope images of foamed metal materials at different magnifications. (**a**) Pre-experiment (50×). (**b**) Post-experiment (50×). (**c**) Pre-experiment (500×). (**d**) Post-experiment (500×). (**e**) Pre-experiment (2000×). (**f**) Post-experiment (2000×).

4. Conclusions

This experiment used a self-designed explosion tube network experimental platform to study the energy absorption performance of three different fiber core layer blast-resistant composite materials on methane–air mixture gas explosions. The energy absorption effect of blast-resistant composite materials with flexible fiber core layers is more significant than that of single foam metal structures. At the same time, the composite material using glass fiber as the core layer has stronger energy absorption performance than the composite materials using carbon fiber and aramid fiber and can provide better protection for the target. In short, by selecting different flexible core layer materials for research, it is helpful to reduce the production cost and reduce the overall weight of the material while ensuring that the overall energy absorption performance of the blast-resistant material is further improved. We believe that our research can not only provide experimental verification for the improvement of the energy absorption performance of foam metal structure blast-resistant materials, but also provide experimental solutions and theoretical analysis for the measurement of energy absorption performance. This study provides enlightenment for further optimizing the energy absorption performance and application of foam metal structures.

5. Patents

Patent application for research-related content: a kind of high-efficiency explosion-proof wall composite energy-absorbing and diffusion-resistant material, No. 202310992747.9.

Author Contributions: Conceptualization, J.T. and B.Z.; Methodology, J.T. and J.C.; Writing—Original Draft Preparation, J.T. and J.C.; Writing—Review and Editing, Y.Z. (Yiyu Zhang), Y.W., Y.H., M.S. and Y.Z. (Yingxin Zhang); Supervision, Y.H. and M.S. All authors have read and agreed to the published version of the manuscript.

Funding: This work was funded by the Heilongjiang Provincial Key R&D Programme Project (No. GA21C023) and by the Basic Research Operating Costs of Undergraduate Colleges and Universities in Heilongjiang Province Project (No. 2020-KYYWF-0680).

Institutional Review Board Statement: Not applicable.

Informed Consent Statement: Not applicable.

Data Availability Statement: Data are contained within the article.

Conflicts of Interest: The authors declare no conflicts of interest.

References

1. Luo, Z.M.; Su, B.; Wang, T.; Cheng, F.M.; Zhao, J.Y.; Wang, L. Research Progress on Explosion Control Technology and Materials of Mining Gas. *J. Saf. Sci. Technol.* **2019**, *15*, 17–24.
2. Zhuang, C.; Wang, Z.; Zhang, K.; Lu, Y.; Shao, J.; Dou, Z. Explosion Suppression of Porous Materials in A Pipe-Connected Spherical Vessel. *J. Loss Prev. Process Ind.* **2020**, *65*, 104–106. [CrossRef]
3. Sharma, V.; Grujovic, N.; Zivic, F.; Slavkovic, V. Influence of Porosity on the Mechanical Behavior During Uniaxial Compressive Testing on Voronoi-Based Open-Cell Aluminium Foam. *Materials* **2019**, *12*, 1041–1053. [CrossRef] [PubMed]
4. Wang, Y.J.; Jiang, S.G.; Wu, Z.Y.; Shao, H.; Wang, K.; Wang, L. Study on the Inhibition Influence on Gas Explosions by Metal Foam Based on Its Density and Coal Dust. *J. Loss Prev. Process Ind.* **2018**, *56*, 451–457. [CrossRef]
5. Wei, C.R.; Zhang, B.R.; Sun, J.H.; Jiang, T.W.; Shi, Q. Experimental Study on Obstacles Affected to Foam Metal to Inhibit Gas Explosion. *Coal Sci. Technol.* **2018**, *46*, 134–141.
6. Yu, M.G.; Liu, M.R.; Wen, X.P.; Pei, P. Study on Synergistic Inhibition of Gas Explosion by Ultrafine Water Mistporous Materials. *J. China Coal Soc.* **2019**, *44*, 1562–1569.
7. Zhang, B.Y.; Cui, J.R.; Tao, J.; Wang, Y.J.; Qin, Y.F.; Wei, C.R.; Zhang, Y.X. Experimental Study on Barrier Performances of Foamed Metals with Different Blast Front Structures to Prevent Methane Explosion Propagation. *Explos. Shock Waves* **2023**, *43*, 170–180.
8. Mudassir, M.; Tarlochan, F.; Mansour, M.A. Nature-Inspired Cellular Structure Design for Electric Vehicle Battery Compartment: Application to Crashworthiness. *Appl. Sci.* **2020**, *10*, 4532. [CrossRef]
9. Alqwasmi, N.; Tarlochan, F.; Alkhatib, S.E. Study of mild steel sandwich structure energy absorption performance subjected to localized impulsive loading. *Materials* **2020**, *13*, 670. [CrossRef]
10. Alkhatib, S.E.; Tarlochan, F.; Eyvazian, A. Collapse behavior of thin-walled corrugated tapered tubes. *Eng. Struct.* **2017**, *150*, 674–692. [CrossRef]
11. Tarlochan, F.; Samer, F. Design of thin wall structures for energy absorption applications: Design for crash injuries mitigation using magnesium alloy. *Int. J. Res. Eng. Technol.* **2013**, *2*, 24–36.
12. Dastjerdi, A.A.; Shahsavari, H.; Eyvazian, A.; Tarlochan, F. Crushing analysis and multi-objective optimization of different length bi-thin walled cylindrical structures under axial impact loading. *Eng. Optim.* **2019**, *51*, 1884–1901. [CrossRef]
13. Gibson, L.J.; Ashby, M.F. *Cellular Solids: Structure and Properties*; Cambridge University Press: Cambridge, UK, 1999.
14. Arora, H.; Hooper, P.A.; Dear, J.P. The effects of air and underwater blast on composite sandwich panels and tubular laminate structures. *Exp. Mech.* **2012**, *52*, 59–81. [CrossRef]
15. Schiffer, A.; Tagarielli, V.L. Underwater blast loading of water-backed sandwich plates with elastic cores: Theoretical modelling and simulations. *Int. J. Impact Eng.* **2017**, *102*, 62–73. [CrossRef]
16. Hoo Fatt, M.S.; Sirivolu, D. Blast response of double curvature, composite sandwich shallow shells. *Eng. Struct.* **2015**, *100*, 696–706. [CrossRef]
17. Rolfe, E.; Quinn, R.; Sancho, A.; Kaboglu, C.; Johnson, A.; Liu, H.; Hooper, P.A.; Dear, J.P.; Arora, H. Blast resilience of composite sandwich panels with hybrid glass-fibre and carbon-fibre skins. *Multiscale Multidiscip. Model. Exp. Des.* **2018**, *1*, 197–210. [CrossRef]
18. Tarlochan, F. Sandwich Structures for Energy Absorption Applications: A Review. *Materials* **2021**, *14*, 4731. [CrossRef] [PubMed]
19. Epasto, G.; Distefano, F.; Gu, L.; Mozafari, H.; Linul, E. Design and optimization of Metallic Foam Shell protective device against flying ballast impact damage in railway axles. *Mater. Des.* **2020**, *196*, 109120. [CrossRef]
20. Dirgantara, T.; Jusuf, A.; Kurniati, E.O.; Gunawan, L.; Putra, I.S. Crashworthiness analysis of foam-filled square column considering strain rate effect of the foam. *Thin-Walled Struct.* **2018**, *129*, 365–380. [CrossRef]
21. Nurick, G.N.; Langdon, G.S.; Chi, Y.; Jacob, N. Behaviour of sandwich panels subjected to intense air blast—Part 1: Experiments. *Compos. Struct.* **2009**, *91*, 433–441. [CrossRef]
22. Karagiozova, D.; Nurick, G.N.; Langdon, G.S. Behaviour of sandwich panels subject to intense air blasts—Part 2: Numerical simulation. *Compos. Struct.* **2009**, *91*, 442–450. [CrossRef]
23. Theobald, M.D.; Langdon, G.S.; Nurick, G.N.; Pillay, S.; Heyns, A.; Merrett, R.P. Large inelastic response of unbonded metallic foam and honeycomb core sandwich panels to blast loading. *Compos. Struct.* **2010**, *92*, 2465–2475. [CrossRef]
24. Buitrago, B.L.; Santiuste, C.; Sánchez-Sáez, S.; Barbero, E.; Navarro, C. Modelling of composite sandwich structures with honeycomb core subjected to high-velocity impact. *Compos. Struct.* **2010**, *92*, 2090–2096. [CrossRef]
25. Roudbeneh, F.H.; Liaghat, G.; Sabouri, H.; Hadavinia, H. High-velocity impact loading in honeycomb sandwich panels reinforced with polymer foam: A numerical approach study. *Iran. Polym. J.* **2020**, *29*, 707–721. [CrossRef]
26. Ha, N.S.; Lu, G.; Xiang, X. Energy absorption of a bio-inspired honeycomb sandwich panel. *J. Mater. Sci.* **2019**, *54*, 6286–6300. [CrossRef]
27. Dharmasena, K.P.; Wadley, H.N.; Williams, K.; Xue, Z.; Hutchinson, J.W. Response of metallic pyramidal lattice core sandwich panels to high intensity impulsive loading in air. *Int. J. Impact Eng.* **2011**, *38*, 275–289. [CrossRef]
28. Beharic, A.; Egui, R.R.; Yang, L. Drop-weight impact characteristics of additively manufactured sandwich structures with different cellular designs. *Mater. Des.* **2018**, *145*, 122–134. [CrossRef]
29. Schenk, M.; Guest, S.D.; McShane, G.J. Novel stacked folded cores for blast-resistant sandwich beams. *Int. J. Solids Struct.* **2014**, *51*, 4196–4214. [CrossRef]

30. Pydah, A.; Batra, R.C. Crush dynamics and transient deformations of elastic-plastic Miuraori core sandwich plates. *Thin-Walled Struct.* **2017**, *115*, 311–322. [CrossRef]
31. Kilchert, S.; Johnson, A.F.; Voggenreiter, H. Modelling the impact behaviour of sandwich structures with folded composite cores. *Compos. Part A Appl. Sci. Manuf.* **2014**, *57*, 16–26. [CrossRef]
32. Zunjarrao, K. A review: Advanced structural and multi-functional sandwich composites with prismatic and foam cores. *Polym. Compos.* **2023**, *pc*, 27849.
33. Mao, X.H. Dynamic Response of the Compound Structure of Foam Aluminum Core Composite Sandwich Material to the Shock Wave of Gas Explosion. *Adv. Mater. Res.* **2013**, *690–693*, 1149–1157. [CrossRef]
34. Zhang, J.; Qin, Q.; Wang, T.J. Compressive Strengths and Dynamic Response of Corrugated Metal Sandwich Plates with Unfilled and Foam-Filled Sinusoidal Plate Cores. *Acta Mech.* **2013**, *224*, 759–775. [CrossRef]
35. Chen, W.; Hao, H.; Chen, S.; Hernandez, F. Performance of Composite Structural Insulated Panel with Metal Skin Subjected to Blast Loading. *Mater. Des.* **2015**, *84*, 194–203. [CrossRef]
36. Zhou, H.; Ma, G.; Li, J.; Zhao, Z. Design of Metal Foam Cladding Subjected to Close-Range Blast. *J. Perform. Constr. Facil.* **2015**, *29*, 04014110. [CrossRef]
37. Langdon, G.S.; Guan, Z.; Cantwell, W.J. Blast Protection for Polymer Composite Materials in Structures Subjected to Air-Blast Loading. In *Comprehensive Composite Materials II*; Elsevier: Amsterdam, The Netherlands, 2018; pp. 332–350.
38. Zhou, N.; Wang, J.; Jiang, D.; Tang, K.; Fang, Y. Study on the Failure Mode of A Sandwich Composite Structure Under the Combined Actions of Explosion Shock Wave and Fragments. *Mater. Des.* **2020**, *196*, 109166. [CrossRef]
39. Santosa, S.P.; Arifurrahman, F.; Izzudin, M.H.; Widagdo, D.; Gunawan, L. Response Analysis of Blast Impact Loading of Metal-foam Sandwich Panels. *Procedia Eng.* **2017**, *173*, 495–502. [CrossRef]
40. Guo, H.; Zhang, J. Expansion of Sandwich Tubes with Metal Foam Core Under Axial Compression. *J. Appl. Mech.* **2023**, *90*, 051008. [CrossRef]
41. Zhang, B.Y.; Tao, J.; Cui, J.R.; Zhang, Y.Y.; Wang, Y.J.; Han, Y.H.; Sun, M. Absorption Characteristics of Methane-Air Mixture Explosion Energy by Foam Metal with A Corrugated Surface Against Explosion. *Explos. Shock Waves* **2023**, *43*, 168–179.
42. Duan, Y.; Long, F.; Huang, J.; Jia, H.; Bu, Y.; Yu, S. Effects of Porous Materials with Different Thickness and Obstacle Layout on Methane/Hydrogen Mixture Explosion with Low Hydrogen Ratio. *Int. J. Hydrogen Energy* **2022**, *47*, 27237–27249. [CrossRef]

materials

MDPI

Article

Friction Investigation of Closed-Cell Aluminium Foam during Radial-Constrained Test

Jozsef Kertesz [1,2,*] and Tünde Anna Kovacs [3]

[1] Department of Vehicle Engineering, Faculty of Engineering, University of Debrecen, 4032 Debrecen, Hungary
[2] Doctoral School on Safety and Security Science, Obuda University, 1034 Budapest, Hungary
[3] Banki Donat Faculty of Mechanical and Safety Engineering, Obuda University, 1034 Budapest, Hungary; kovacs.tunde@bgk.uni-obuda.hu
* Correspondence: kertesz.jozsef@eng.unideb.hu

Abstract: The energy-absorbing capacity and friction phenomena of different closed-cell aluminium foam-filled Al tube types are investigated through experimental compression tests. Concerning the kind of investigation, free, radial-constrained and friction tests occurred. The radial-constrained compression test results confirm that the process requires significantly more compression energy than without the constrain. Pushing away different pre-compressed foams inside the aluminium tube, the static and kinematic frictional resistances can be determined and the energy required to move them can be calculated. Knowing the value of the energy required for the frictional resistance, we can obtain how much of the energy surplus in radially inhibited compression is caused by the friction phenomena. The main goal present study is to reveal the magnitude of friction between the foam and the wall of the tube during the radially constrained test. The investigation used 0.4 and 0.7 g/cm^3 density closed-cell aluminium foam whilst a compressive test was applied where the force–displacement data were recorded to calculate the absorbed energy due to friction. Considering the results of the test, it can be stated that 18% of the invested energy was used to overcome friction in the case of lighter foam and almost 23% with 0.7 g/cm^3 foam during the radial-constrained test.

Keywords: friction; closed-cell aluminium foam; compression test; radial constrain; absorbed energy; crashworthiness

Citation: Kertesz, J.; Kovacs, T.A. Friction Investigation of Closed-Cell Aluminium Foam during Radial-Constrained Test. *Materials* **2024**, *17*, 3344. https://doi.org/10.3390/ma17133344

Academic Editors: Yuyuan Zhao and Huiping Tang

Received: 14 May 2024
Revised: 2 July 2024
Accepted: 3 July 2024
Published: 5 July 2024

1. Introduction

Aluminium foams with closed cells are used in structural applications and provide lightweight and impact-absorbing qualities for automobiles [1–4]. These foam's mechanical characteristics are shaped by the way their topology and geometry change upon impact. Made with powder metallurgy, these materials have significant energy absorption capacities and are deemed promising for use in a range of sectors [5–7]. According to research, vehicle crash boxes filled with aluminium foam have far better energy absorption per unit mass and length, allowing for shorter crash boxes with the same amount of energy absorption [8–11]. Through the creation of a lightweight, modular composite framework that can be adjusted for impact absorption, this technique improves structural safety in automobiles [12]. Research has investigated the application of foam-filled aluminium tubes in automobile production to improve crashworthiness and maximise lightweight design. Pore structure, cell wall flaws, and cell wall microstructure are some of the elements that affect the mechanical characteristics of closed-cell aluminium foams [13–15]. The energy absorption capacity of closed-cell aluminium foam remains almost constant throughout a range of strain rates, but the compressive behaviour exhibits notable strain rate hardening [16,17]. When compared to other cellular metals, the specific energy absorption of closed-cell aluminium foam demonstrates a fair degree of agreement in terms of deformation energy. The mechanical response of closed-cell aluminium foam can now be predicted quickly, reliably, and accurately thanks to the development of a computer model that has been verified by experimental data [18–21]. The mechanical properties of closed-cell aluminium foams have been extensively studied in both confined and unconstrained situations [22–24]. Additionally, the same foams have been tested and placed into tubes. From the point of view of crashworthiness safety development, the main goal of the investigation is to maximalise the energy absorbing capacity of the actual foam, without significant weight increase [25,26]. One of the most improved ways to reveal the absorbing behaviour of the foam is the radial-constrained compression condition. This means the foam is inserted into a thin-walled structure as, for example, the crash box is, and this porous material is compressed inside the inhibitor. The inhibitor could be made from different materials such as aluminium tubes or plates, but the composite is also applied as a constraint [27–29]. However, the relationship between the energy surplus and the friction phenomenon occurring during the radial force test is not detailed in the literature.

Due to the surface complexity of a closed-cell foam, it is hard to define its surface roughness and its friction conditions. The size of the cells has an effect not only on the energy-absorbing capacity but also on the friction between the foam and the connected surface [30]. One important component of material behaviour that affects the functionality and uses of closed-cell aluminium foams is friction [31–34]. However, there was no relevant information about friction inside of the Al tube during the radial-constrained compression test. Therefore, the main goal of the present research was to reveal the magnitude of Coulomb friction during a radial-constrained test of closed-cell aluminium foam, for which a complex friction measurement method was used. This study is part of more complex research, namely the enhancement of vehicle crash safety with structural solutions, where the goal is to create a multi-membered crash box structure filled with metal foam. The results detailed here significantly help our further development work, and hopefully provide useful and inspiring information on the science of crashworthiness with metal foam.

2. Materials and Experimental Methods

2.1. Preparation of Specimens

The present research focuses on the energy-absorbing behaviour of closed-cell aluminium foam to which the base material was manufactured by Behai Composite Materials Co. (Jiujiang, China) in the form of 600 × 600 × 30 mm blocks with two different densities, numerically 0.4 and 0.7 g/cm^3 [35]. The chemical ingredients of the foam were as follows: 97% aluminium powder, 3% foaming agent, 1% TiH_2, increasing viscose, and 2% Ca. They are environmental protection elements that have no pollution. For the compressive test, 30 mm diameter and height specimens were made using a waterjet cutting process. Owing to the precise cutting process, the specimens had a consistent height and diameter which were measured using a digital calliper at more points along the foam's height. To not influence the test results by the possible inner material or structure issues of foam specimens, all of the measurements were repeated three times to increase the statistical meaning, but the later results are around the mean of values. During the radially constrained investigation, the foam specimens were inserted into a 30 mm inner diameter aluminium tube with a wall thickness of 5 mm and 65 mm height and compressed using a pushing plate with a 28 mm nominal diameter and 3 mm thickness. The tube preparation did not require any type of surface finishing or machining over the cutting up with 65 mm. Considering the material of the tube and pushing plate, the AlMgSi AW EN6061 was applied as one of the frequently used materials in the field of automotive construction, mainly in the bumper system and its accessories. Between the test machine crosshead and the pushing plate, a so-called push rod was inserted in the form of a 28 mm nominal diameter tube. The prepared specimens before the test are presented in Figure 1, whilst their geometry data are collected in Table 1. According to the dimensions and weights, the theoretical densities are confirmed.

Figure 1. Specimens for the test.

Table 1. Dimensions of specimens.

Designation of Specimens	Lighter Foam			Denser Foam		
	Mass [g]	Volume [cm^3]	Density [g/cm^3]	Mass [g]	Volume [cm^3]	Density [g/cm^3]
SP1	9.24	21.21	0.43	14.25	21.19	0.67
SP2	9.20	21.17	0.43	14.05	21.19	0.66
SP3	9.31	21.09	0.44	14.31	21.32	0.67
Average	9.25	21.15	0.43	14.20	21.26	0.67

2.2. Test Conditions and Tools

For the test, a INSTRON 8801 servo hydraulics material fatigue test machine was used, connected with a data collector and analysis software, namely WaveMatrix v.2, to record the force and displacement values. The velocity of the crosshead of the test machine was constant at 1 mm/s, whilst the system data recording frequency was 1000/s. The displacement is recorded using an implement displacement encoder, and the force is detected using a load cell with a limit of 100 kN. The measurement was carried out in an accredited laboratory and the measuring instrument had a valid calibration and verification logbook at the time of measurement. Before the test, owing to the prevention of technical integrity of the load cell, the 80 kN maximum force limit was adjusted; if this condition was exceeded, the operation was immediately blocked. The components of the test machine are in Figure 2.

2.3. Scheme of Compressive Tests

The main goal of the present investigation is to reveal the effect of friction concerning the radial constraint between the foam and tube compared with the uniaxial (without radial constraint) test. First, the foam was loaded using the crosshead without any radial inhibitor, resulting in a uniaxial compressive test. As a continuation of the test, the foams were inserted into the above-mentioned aluminium tube. The scheme of the radial constrained version can be seen in Figure 3.

The nominal diameter of foam specimens is equal to the inner nominal diameter of the tube and pushing plate, but to reveal the type of fit between them the real dimensions must be considered. Taking into account the real diameters, the transition fit occurred between the foam and the tube; so, to make this fit, a slight assembling force had to be applied on the foam. Concerning the fit of the pushing plate and tube clearance, the type of fit was applied to not influence the friction measurement between the foam and wall of the tube during the compression test. In the case of the pushrod, the type of fit does not make sense, since the nominal size of them is not the same. In order not to influence the measurement results, no lubricant was used for the assembly or during the compression test.

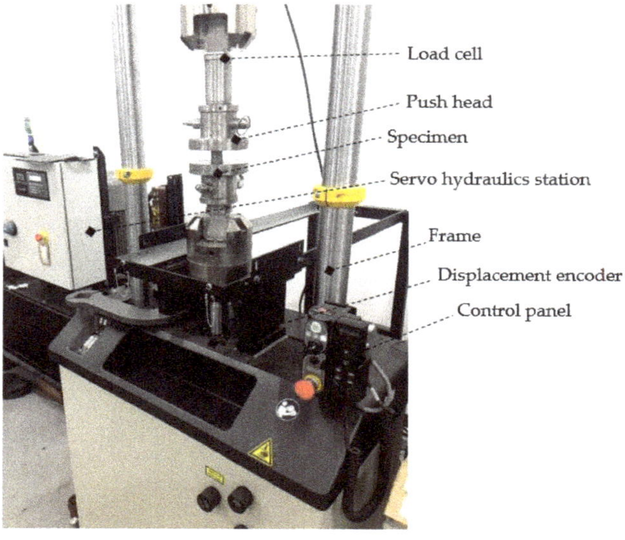

Figure 2. Tools for investigation.

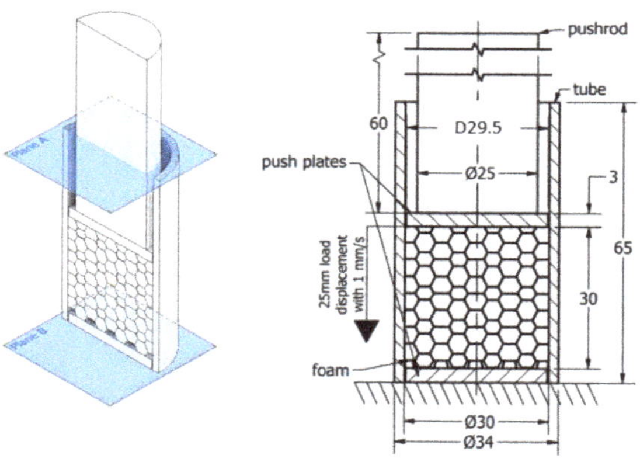

Figure 3. The scheme of radial-constrained test (Plane-A and B help to understand the orientions).

3. Radial-Constrained and Friction Test

3.1. Steps of Radial-Constrained Test

During the test, the foam is loaded using axial compressive force inside the aluminium tube. Considering the Poisson effect of materials, the compression or tensile load is accompanied with radial strain [36]. In this case, the compression caused by radial strain is inhibited by the wall of the tube. Considering the closed foam structure as a 2D hinged structure, the DoF of its lateral node is blocked in the direction of the y-axis, increasing the stiffness and strength of construction. The simplified structures with nodes can be seen in Figure 4, where F_{y}^{c}. is the force of the compressive load by the node, F_{y}^{s}. is the supporting or reaction force, F_N is the radial normal force, whilst F_{μ} represents the friction force.

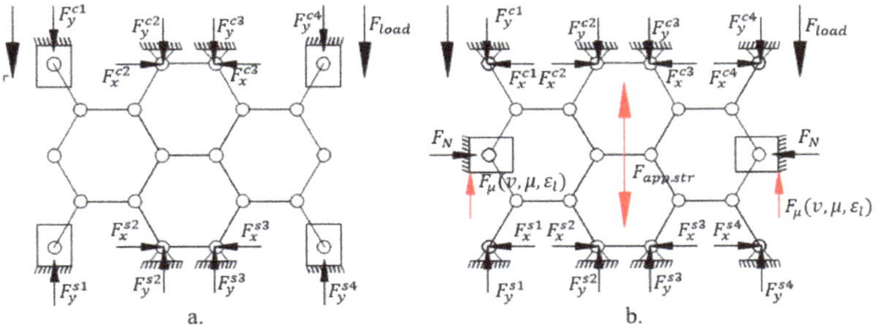

Figure 4. Representation of hinged structure. (**a**) free compressive, (**b**) radial-constrained.

However, it must be taken into account as an apparent stiffness enhancement caused by the presence of the wall of the tube and the inhibition of radial size increase. Owing to the inhibition of lateral displacement, the load results in cell wall deformation require more energy during the axial displacement. The cell wall can be matched with as a bar of the 2D hinged structure. Due to the presence of friction resistance between lateral nodes and the wall of constraint, the compression requires even more energy. Furthermore, considering the Poisson effect and the scale of the contact surface between the foam and tube, the conditions of friction resistance change continuously, affecting the required load. It was important to reveal the volume of the energy requirement concerning overcoming this friction resistance during the radial constrained test.

3.2. Method of Friction Test

The friction investigation could be divided into three main stages. In the case of the first stage, both density foams (0.4 g/cm^3 and 0.7 g/cm^3) were loaded individually via compression with 25 mm crosshead displacement, namely, a free compression test occurred. The test was applied three times. In the follow-up, the foam specimens were inserted into a tube, and the load was placed inside the tube using the pushing rod and plate. The applied displacement was the same, numerically 25 mm. Comparing the results of these two tests, the absorbed energy surplus of the radial-constrained version can be expressed easily.

To obtain the friction-caused energy proportion of this absorbed energy surplus, the third test stage is required. This process contains more sub-steps that continuously modify the constraints of the structure. Considering the radially constrained test seen in Figure 3, five different sections were made using different strains concerning the foam. In the case of the first specimen, the foam was inserted into the tube and non-loaded with any compression force. In the case of the second specimen, the foam is compressed with a 5 mm displacement, resulting in a 25 mm height. The third specimen is compressed with a 10 mm displacement to obtain a 20 mm height. The fourth and fifth ones are compressed with 15 and 20 mm, respectively. After the compression loads, the pre-compressed specimens are rotated and, using a push plate and pushrod, the foam is pushed away inside the tubes whilst the displacement is recorded to obtain information about the friction resistance.

Considering the five different compression states—owing to the Poisson effect—the radial/normal force (P_{max}) between the tube wall and the foam gets higher and higher whilst the area of the contact surface is modified, influencing the friction resistance during the foam displacement inside the tube. The concepts are introduced in Figure 5. The meaning of the colours is next. The green vectors describe the radial normal force (P_{max}) generated by the Poisson effect. The direction of vectors shows the direction of radial force loading the wall of the tube, whilst the length of them represents the magnitude of this normal force. The red lines indicate the friction force which is influenced by the P_{max} and friction coefficient (μ). The blue hatched area symbolises the final position of the pre-compressed foam after they were pushed away inside the tube with a 30 mm displacement.

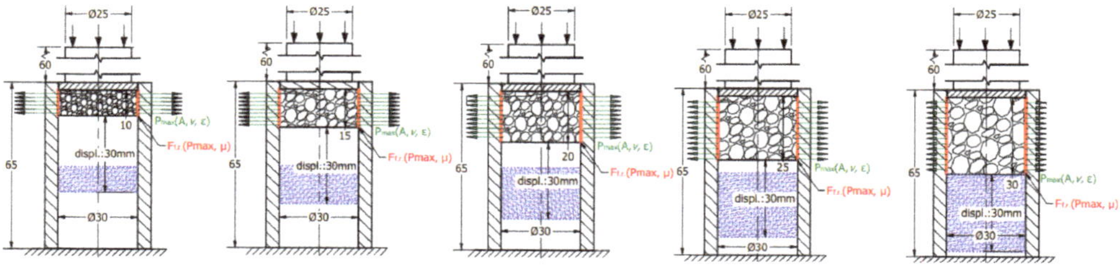

Figure 5. Scheme of friction measurement of pre-compressed foams.

4. Friction and Contact Surface Issue

4.1. Friction Phenomena

Friction is one of the most interesting resistances in the field of engineering since sometimes it must be avoided, but sometimes, without its presence, the process cannot occur [37,38]. The friction is the resistance of the movement that occurs between the objects. The values strongly depend on more parameters like surface roughness, normal force, the speed difference between the contact surfaces, and the presence or lack of the lubricator. The friction force can be expressed using the well-known Coulomb Formula (1) [39,40].

$$F_{fric.} = \mu \cdot F_N \tag{1}$$

where μ is the friction coefficient, whilst F_N is the normal force between the connected objects. Physics fundamentals include the ideas of static and kinetic friction, and during the present test, both of them must be considered. Kinetic friction happens between surfaces that are moving, whereas static friction is the force that stops an item from moving when a force is applied [41–43]. While the coefficient of kinetic friction affects kinetic friction, the coefficient of static friction determines the amount of static friction concerning the normal force. Static friction often outweighs kinetic friction. The theoretical graph of the friction issue is introduced in Figure 6.

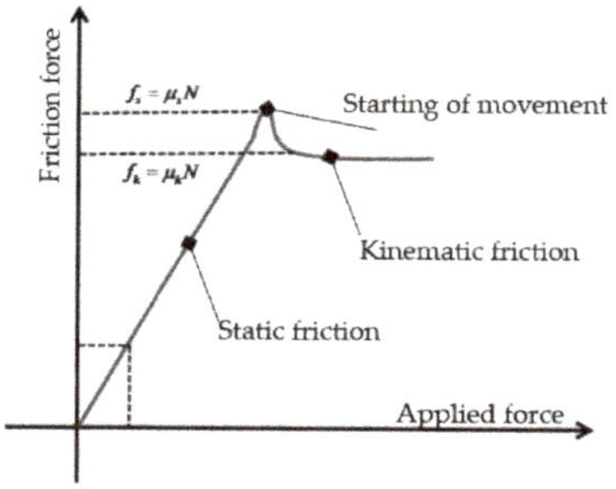

Figure 6. Graph of theoretical friction force in function of applied force.

4.2. Contact Surface Calculation

According to the science of friction, the contact surface size is not a crucial parameter during the investigation of friction issues. According to Figure 5, in this case, the normal force (P_{max}) and the size of the contact area are also changed continuously, affecting the friction resistance and influencing the energy requirement for the movement of pre-compressed foam. Due to the porous structure and surface of metal foams, it is not easy to define the exact area of contact surfaces of specimens inside the tube. To be informed about the scale of the contact area of cylindrical specimens, we performed microscopic analysis. For the image acquisition, the Zeiss Smartzoom 5 smart digital microscope was used with its own official Zeiss Zen v3.8 core software, presented in Figure 7.

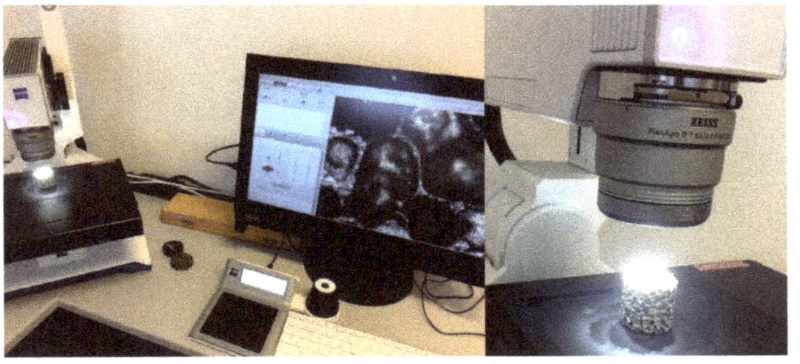

Figure 7. Microscope investigation to reveal the size of void inclusions and their deviations.

Using the built-in digital measurement tools of the microscope, it is easy to measure the size of void inclusions and calculate the average distribution density. The print screen of the microscopic view with the dimension measuring tool is in Figure 8. Both density-type foams were subjects of the microscope measurements and presented the standard deviation in the Gauss diagram, whose results can be seen in Figure 9. The average void inclusions are 3680 μm and 2570 μm in the case of 0.4 g/cm^3 and 0.7 g/cm^3 density foam, respectively.

Figure 8. Microscopic view of the surface of foam specimen before the test.

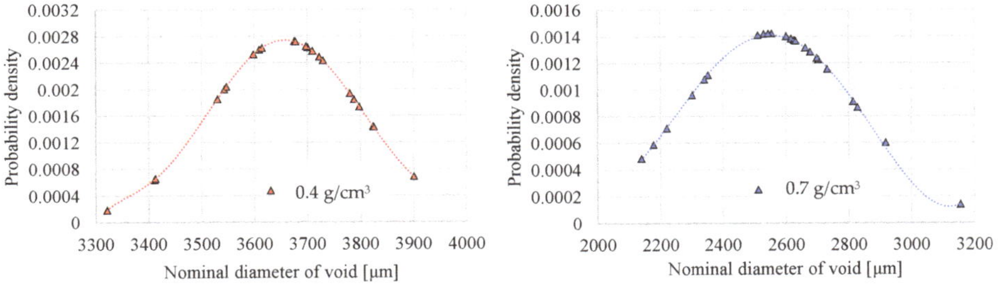

Figure 9. The Gauss curve around size of void inclusions of each foam.

To reveal the size of the contact surface between the foam and tube wall, a 100 mm^2 unit area was first taken into account, where the number of cells can be easily to counted using a digital microscope and then extended for the whole surface of the cylindrical specimen, considering the aforementioned and detailed standard deviations. According to this method, 14.6 cells can be accounted for per unit of 100 mm^2 area in the case of 0.7 g/cm^3 type foam, and 8.3 cells in the case of 0.4 g/cm^3 foam. Applying this calculation to the entire casing of the cylindrical test specimen, the contact area is the sum of the wall thicknesses between the inclusions on the surface. In the case of the 30 mm in diameter and height cylindrical solid specimens, they would have a 9.42 cm^2 surface. Considering the number, size and deviation of void inclusions, the outer contact surface of foam is 2.42 cm^2

in the case of denser foam, and 1.13 cm^2 in the case of 0.4 g/cm^3 type foam. These values are considered in the case of foams before the loading. Compressing the foam results in more and more cell deformation involving contact surface modification. The representation of this contact surface issue by the 0.4 and 0.7 g/cm^3 foam is introduced in Figure 10, painted in red.

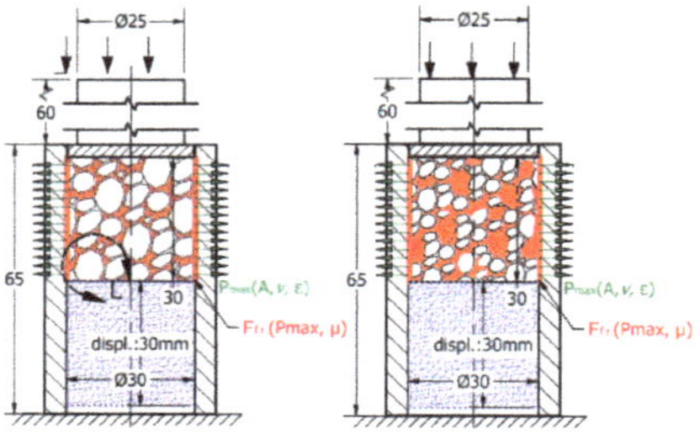

Figure 10. Representation of contact surface with different density foam.

5. Evaluation Criteria

More evaluation criteria were used during the investigation to analyse the effect of the radial constraint and the friction between the foam and tube. The present study is a part of vehicle crashworthiness safety development; therefore, the absorbed energy by the foam is a focus area, as used for evaluation criteria. To obtain the absorbed energy by the specimens, the area under the force–displacement curve must be considered using graph integration (2), with Joules as the units of the values [44–46]. Concerning the closed-cell aluminium foam force–displacement test, the diagram can be divided into three main sections, namely the linear elastic section, plateau stage and densification zone [47–50]. Referring to this period, the absorbed energy is determined in the function of displacement to reveal the absorbing behaviour of each section as well.

$$EA = \int F(x)dx\ [J] \tag{2}$$

From the point of view of vehicle crashworthiness safety development, weight optimisation is another crucial requirement; therefore, *SEA*, defined as Special Energy Absorbing, occurred, as one of the most frequently applied evaluation criteria in this research field (3) [51,52]. To determine this value, the ratio of absorbed energy and mass of specimens was calculated. However, according to another view, not only is the mass a crucial criterion, but the volume of energy absorber components of the crumple zone is also. Naturally, a larger crumple zone could involve higher absorbing capacity, but this scale is strongly limited by the required aesthetic design and shape. Unduly enlarged crumple elements take space away from other important components such as coolers, inverters, and control units, and can significantly reduce the car's payload storage space [53]. Considering this issue, the definition of Volume Fraction (4) is used as a kind of useful complement or alternative to the frequently used *SAE*.

$$SEA = \frac{\int F(x)dx}{m} \left[\frac{J}{g}\right] \tag{3}$$

$$VF = \frac{\int F(x)dx}{V} \left[\frac{J}{cm^3}\right] \tag{4}$$

where $\int F(x)dx$ is the absorbed energy of the relevant zone, m is the mass, and V is the volume of the foam specimen. Using the Crashing Force Efficiency as an evaluation criterion, the ratio of the first maximum force and the mean value of the plateau zone can be expressed (4). Regarding the interpretation of CFE, the value is accounted in units of percent, and the main goal is to reach higher and higher efficiency [54,55].

$$CFE = \frac{F_{avrg.}^{plateau}}{F_{max}} [\%] \quad (5)$$

where $F_{avrg}^{plateau}$ is the average resistance force of foam during the compression, whilst the F_max is the first maximum force of compression. The closed-cell aluminium foam could be a useful energy absorber in the crumple zones of vehicles, where the consistent absorbing behaviour is one of the most important requirements. Using the Crashing Force Fluctuation is suitable for analysing the assessment compliance with this criterion focusing on the section of plateau. The CFF is accounted as the ratio between the real absorbed energy of the plateau zone and the mean force-based absorbed energy of the plateau zone (6). This evaluation consideration is frequently named ULC—Undulation of Load Carrying Capacity [56–58].

$$CFF = \frac{\int |F(x) - F_m| dx}{\int F(x) dx} \quad (6)$$

where $\int |F(x) - F_m| dx$ is the energy different considering the real and mean of crushing force, while $\int F(x) dx$ is the real absorbed energy.

6. Results of Uniaxial Load

As the first step, the foams were loaded by 1 mm/s without any type of radial inhibitor accessories. The foam graph followed the typical force–displacement diagram with linear elasticity, plateau and densification. To obtain a stress–strain curve, compressive force must be divided by the initial cross-sectional area perpendicular to the loading direction. The effect of foam density can be considered; concerning the first maximum force, lighter (0.4) and denser (0.7) foam have 5.66 kN and 7.94 kN, respectively, as the first local maximum in the force–displacement curve. During the test, the ISO 13314 (2011) standard is taken into consideration, according to which the plateau stress is the arithmetical mean of the stresses at 0.1% or smaller strain intervals between 20% and 30% or 20% and 40% compressive strain [59]. The mean value of plateau stress was less with 15% in the case of lighter foam than the denser one. Numerically, the average stress in the plateau zone was 6.92 MPa concerning lighter ones and 12.26 MPa in the case of denser. The end of the plateau zone is defined by the above-mentioned ISO standard as the point in the stress–strain curve at which the stress is 1.3 times the plateau stress. Considering this rule, it is easy to define the densification point of the specimen, numerically 64.28% and 50.53% regarding the lighter and denser foam, respectively. The graphs of the free compression tests are presented in Figure 11, where the vertical dotted lines indicate the end of plateau.

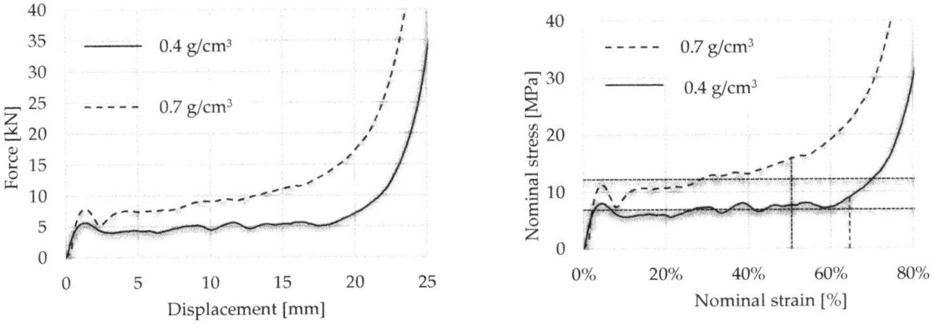

Figure 11. Displacement force and stress–strain curves of free compression test.

7. Results of Radial-Constrained Test

The hypothesis according to which the radially constrained foam results in a higher energy level during its compressive load is confirmed by the present investigation as well. The typical sections such as linear elastic zona, plateau and densification can be observed in this test situation as well. However, the plateau zone has a shorter length and a slight steepness compared with the uniaxial load version. During this compressive load, the force displacement shows higher characteristics, explained by the above-mentioned apparent strength increase and friction. Numerically, the plateau zone in this case can be characterised with 24.48 MPa in the case of denser foam, while with the lighter foam, this value was 9.51 MPa. Another significant deviation between the two concepts was the densification strain. The results of the radially constrained test are in Figure 12 and are compared with the uniaxial load.

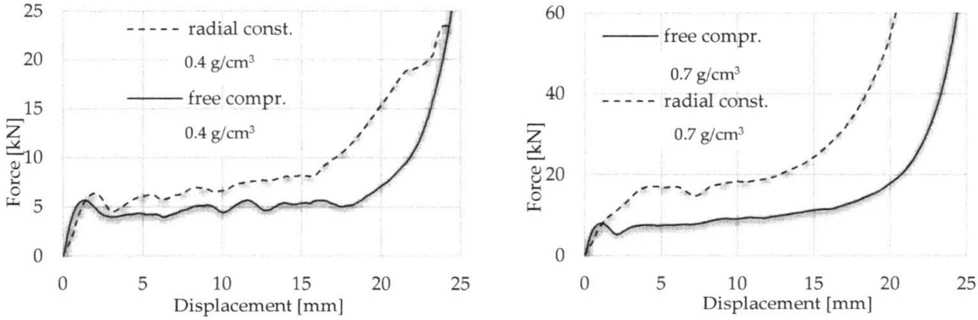

Figure 12. Displacement force diagram to compare the free and radial-constrained compressive tests.

8. Absorbed Energies and Efficiencies

Considering the energy levels, the 0.4 g/cm^3-type foam in the radial-constrained version was able to absorb 32.5% more energy during the compression compared with the uniaxial load concept. The 0.7 g/cm^3 foam called for 90.5% more compression energy during the radial-constrained test. Since the tube wall inhibited the radial deformation of cells of foam specimens, the end of the plateau zone in these cases occurred earlier. Numerically, the 0.4 gm/cm^3 and 0.7 g/cm^3 foam have plateau stress with 7.8 MPa and 14.5 MPa, respectively. Considering the above-mentioned ISO standard, it is easy to define the end of the plateau since it is 1.3 times of plateau value. The end of the plateau is the starting moment of the densification zone, which occurred at a strain of 64.28% and 53.86% in the case of 0.4 g/cm^3 foam during the free and radial-constrained test, respectively. With the same conditions, the appearance of densification of the denser foam was 50.53% and 47.19%. Analysing the crashworthiness behaviour of the absorber, the plateau zone has a dedicated role; therefore, the absorbed energies and efficiencies are calculated concerning the plateau zone as well. The plateau zone and densification issues are collected in Table 2. The SAE and VF values are not relevant for the foam in the case of the radially constrained test, since in that case the weight and volume of the aluminium tube and the accessories ought to be taken into account as well. The results of the present investigation could be also validated by the study of Yang et al. where different density types of closed-cell aluminium foams were tested and confirmed the values of 0.43 and 0.7 foam as well [60].

As it was mentioned above, the loading was limited at 80 kN due to the prevention of the overloading of the load cell; therefore, the radially constrained compression test of the 0.7 g/cm^3 foam was able to run up to a strain of 66.29% which only corresponds to the displacement of 19.88 mm. The energies are collected in Tables 3 and 4; these are presented in Figures 13 and 14.

Table 2. Plateau range relevant results.

Results	Free Compr. Test of 0.4 g/cm^3	Radial Const. Test of 0.4 g/cm^3	Free Compr. Test of 0.7 g/cm^3	Radial Const. Test of 0.7 g/cm^3
Plateau stress (20–40% strain) [MPa]	6.92	9.51	12.26	24.48
Nominal strain of end of plateau [%]	64.28	53.86	50.53	47.19
Stroke of end of plateau [mm]	19.28	16.15	15.16	14.16
EA up to end of plateau zone [J]	93.96	104.33	126.13	225.31
SEA up to end of plateau zone [J/g]	10.15	not relevant	8.88	not relevant
VF up to end of plateau zone [J/cm^3]	13.31	not relevant	17.86	not relevant
CFE up to end of plateau zone [%]	77.91	106.13	108.43	112.54
CFF of plateau zone [%]	0.15	3.8	1.11	29.61

Table 3. Absorbed energy by 0.4 g/cm^3 foam in each period given in Joules.

Stroke [mm]	Free Compression Test	Radial-Constrained Test	Differential	
0–5	20.87 J	23.86 J	2.99 J	+11.4%
0–10	43.48 J	53.37 J	9.89 J	+22.7%
0–15	68.88 J	94.42 J	25.54 J	+37.1%
0–20	97.51 J	149.79 J	52.28 J	+53.6%

Table 4. Absorbed energy by 0.7 g/cm^3 foam in each period given in Joules.

Stroke [mm]	Free Compression Test	Radial-Constrained Test	Differential	
0–5	31.53 J	59.84 J	28.31 J	+89.7%
0–10	72.21 J	142.93 J	70.72 J	+97.93%
0–15	121.54 J	244.03 J	122.49 J	+100.1%
0–20	190.03 J	415.18 J	225.03 J	+118.48%

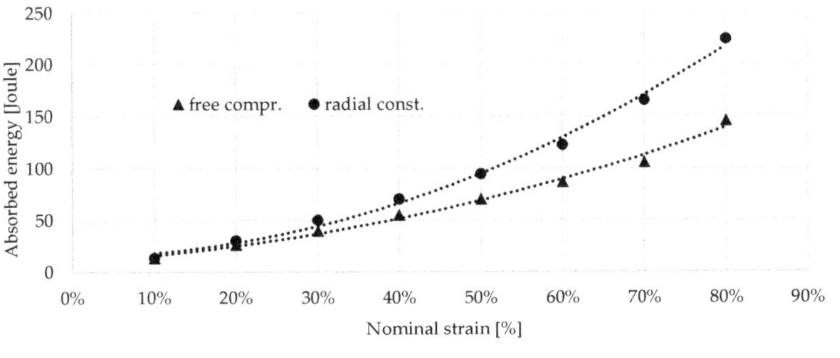

Figure 13. Absorbed energy by 0.4 g/cm^3 foam between 10 and 80% nominal strain.

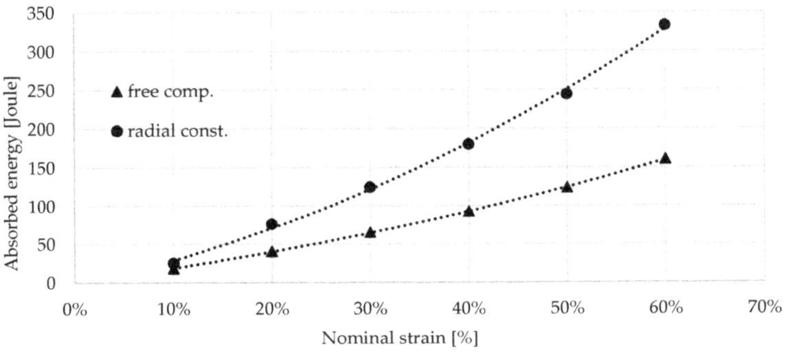

Figure 14. Absorbed energy by 0.7 g/cm^3 foam between 10 and 60% nominal strain.

9. Results of Friction Measurement

As the description above confirmed, the radially constrained condition for foam involves a higher compressive energy requirement due to the inhibited cell deformation and the overcoming of friction between the foam and the wall of the tube. The main goal of this study is to reveal the friction proportion during the radial-constrained load. The steps of the friction measurement are introduced in Section 3.2.

Firstly, the 0.4 g/cm^3 foam was partially loaded by 5 mm compression displacement. Then, the pre-compressed foam was pushed via the tube from the other direction via the tube with 30 mm of displacement whilst the force–displacement curve was recorded to obtain a friction chart about the movement. The measurement was continued by pre-loading another foam specimen with a displacement of 10 mm and then pushing it through the tube from the other direction with a displacement of 30 mm. This step was followed with 15 and 20 mm pre-loading and friction measurements. Naturally, the process was repeated, involving 0.7 g/cm^3 foam as well.

According to the measurement, the graph follows the typical static–kinematic form of a diagram. The results of the test confirmed the initial hypothesis, according to which the notable part of the energy surplus occurred in the radially constrained test caused by friction. Through analysing the graphs, the typical kinematic and static resistance can be obtained. When the 0.4 g/cm^3 foam was pre-compressed with 5 mm displacement, 0.51 kN force was applied to move it in the tube, and then the kinematic friction prevailed and called for a 0.29 kN moving force. Observing these values regarding pre-compressed foam with 10 mm, the kinematic resistance was 0.32 kN, whilst the static value of it was 0.97 kN. In the pre-loaded foam using 15 mm compression, the static resistance required 1.47 kN force, whilst the kinematic result showed 0.69 kN. In the case of the 20 mm-pre-loaded foam, the static resistance was 1.85 kN, and the kinematic resistance was 0.91 kN. And naturally, it is important to know the friction of foam without pre-load. In this condition, the movement required 0.16 kN force to overcome the static friction and 0.06 kN force to keep it moving. Then, the friction investigation steps were repeated, related to the 0.7 g/cm^3 foam as well. As was expected, the friction resistances were significantly higher, since the connection area of specimens is larger. This surface phenomenon is confirmed using the void-area calculation via the smart microscope view detailed in Section 4. The test includes 5–10–15–20 mm pre-compression where the static frictions were 0.87 kN; 7.89 kN; 10.05 kN; and 12.21 kN, respectively. The movement of the non-pre-loaded specimen called for 0.53 kN force to overcome the static friction. The forces required to keep the test specimens in motion to exceed the kinematic friction are, respectively: 0.26 kN; 3.25 kN; 4.53 kN; and 5.77 kN. In the case of the non-pre-loaded version, the foam movement required 0.08 kN. The results of the friction measurement step-by-step are presented in Figure 15, whilst Figures 16 and 17 summarise the graphs of each foam. Table 5 is about the collection of static and kinematic friction resistance in the function of pre-loading specimens. After the pre-compressed foam is pushed away inside the tube, the worn, scratched surface is well-observed. A section view was made about the 15 mm pre-compressed foam after the friction, measuring to show the surface roughness and the final position of the foam. The section view is presented in Figure 18 and is compared with the initial sketch drawing.

Considering the results of the above test, the friction graph of the radial-constrained tests can be constructed using the values of kinematic friction forces and the initial static friction force. These graphs are introduced in Figures 19 and 20. The diagrams start with the first local maximum presenting the overcoming of static friction, then are followed by the kinematic friction resistance, which can be observed. However, due to the inner compression of the foam, the radial force (P_{max}) between the foam and the tube wall increases continuously, which is manifested in an increasing kinematic friction characteristic. The radial compression (P_{max}) increase issue is presented in Figure 5, indicated by green arrows. Using the linear trend line, the friction resistance can be predictable as a function of further displacement. The usefulness of the friction functions is that they can be used to determine how much energy is required to overcome the frictional resistance arising during the radially inhibited compaction. To define this energy, the integration of friction graphs is required, with up to 25 mm displacement; these are then divided into different ranges.

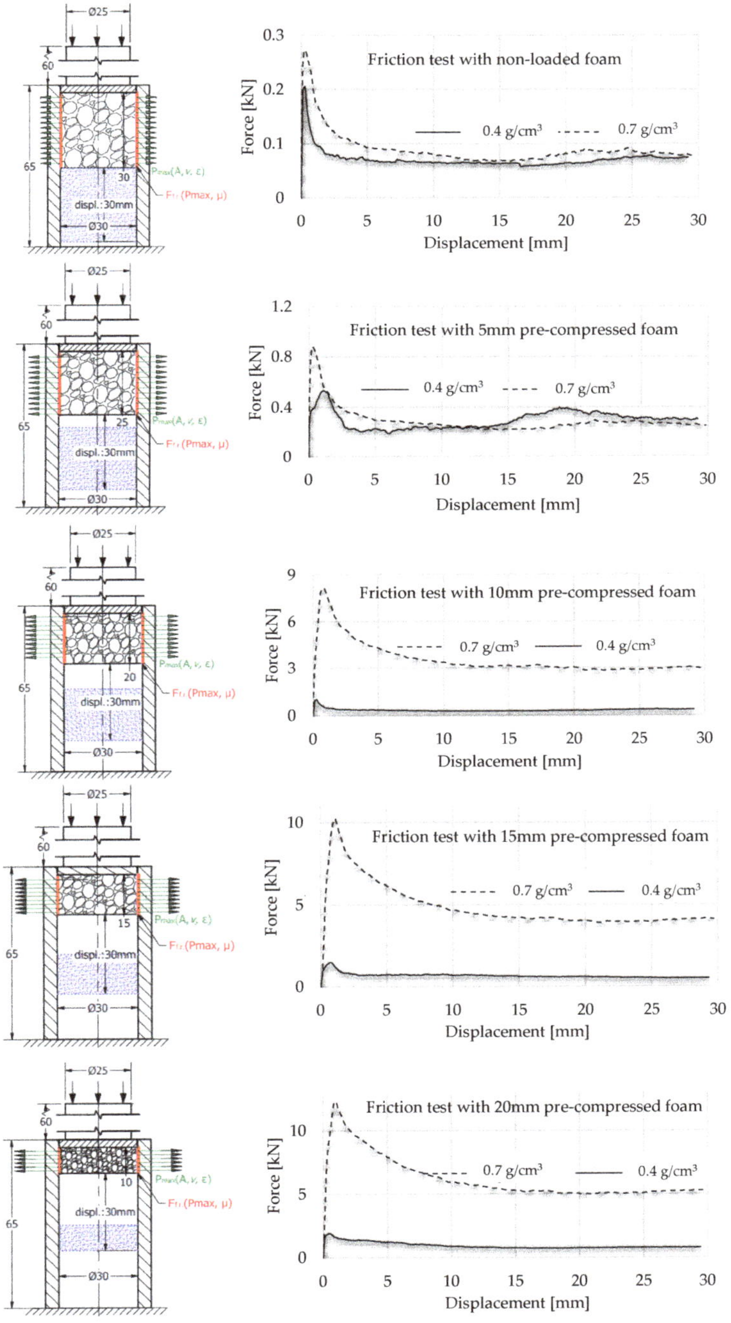

Figure 15. Friction graphs of each pre-compressed specimen pushed away inside the tube.

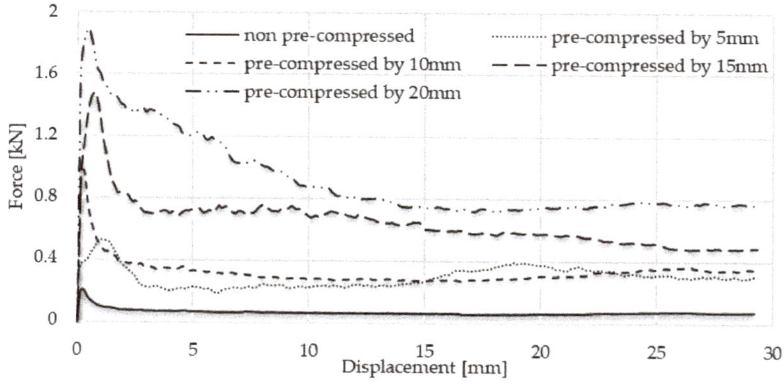

Figure 16. Comparison of friction graphs of 0.4 g/cm^3 foam.

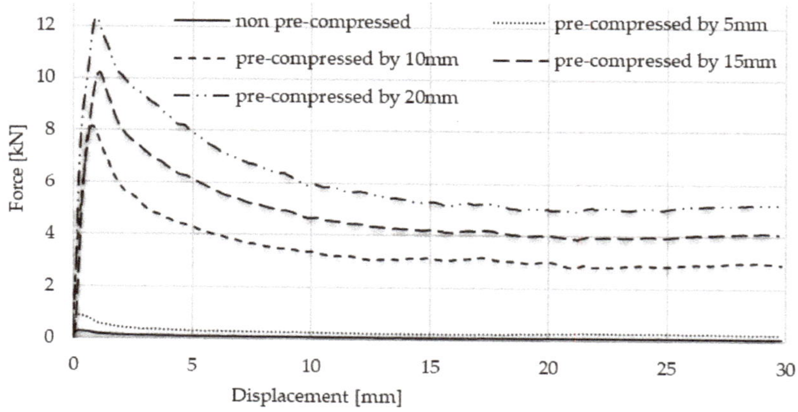

Figure 17. Comparison of friction graphs of 0.7 g/cm^3 foam.

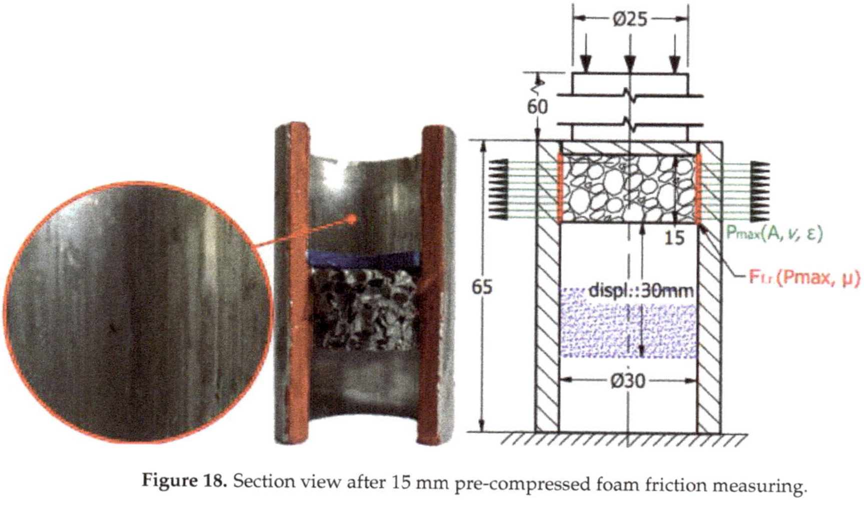

Figure 18. Section view after 15 mm pre-compressed foam friction measuring.

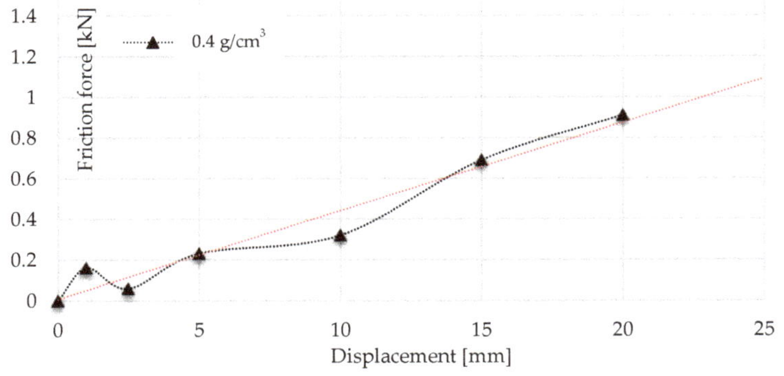

Figure 19. Trend of frictional resistance during multiaxial test using foam with a density of 0.4 g/cm^3 with trendline.

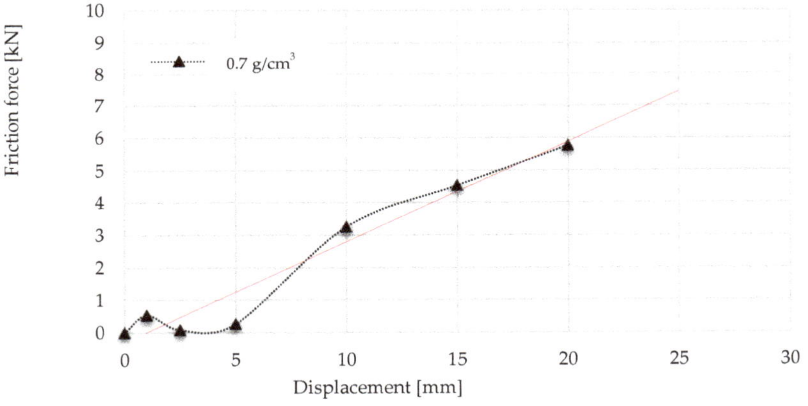

Figure 20. Trend of frictional resistance during multiaxial test using foam with a density of 0.7 g/cm^3 with trendline.

Table 5. Friction resistances of different pre-compressed foam.

Foam Pre-Loading	Mean of Static Friction Force [kN]		Mean of Kinematic Friction Force [kN]	
	0.4 g/cm^3	0.7 g/cm^3	0.4 g/cm^3	0.7 g/cm^3
Non-pre-compressed	0.16	0.53	0.06	0.08
Pre-compressed by 5 mm	0.51	0.87	0.29	0.26
Pre-compressed by 10 mm	0.97	7.89	0.32	3.25
Pre-compressed by 15 mm	1.47	10.05	0.69	4.53
Pre-compressed by 20 mm	1.85	12.21	0.91	5.77

According to the integration of friction graphs, the radial-constrained compression of 0.4 g/cm^3 foam required 0.6 J of energy to overcome the friction up to the displacement of 5 mm. Above 5 mm, due to the increase in radial force (P_{max}) of foam, 1.98 J was needed to overcome the friction resistance. Going on with the displacement, 4.51 J of energy was required to keep moving the object and overcome the kinematic friction between the foam and tube wall. Summarising the energy requirement up to 20 mm, 8.52 J was needed to overcome the friction. In the case of 0.7 g/cm^3 foam, the radially constrained concept required 28.31 J more energy up to the 5 mm displacement. Of this energy surplus, 1.08 J was used to overcome friction, whilst 9.86 J of energy surplus was needed up to 10 mm displacement owing to the presence of resistance. A total of 28.82 J of energy was required in the case of a radial-constrained version of 0.7 g/cm^3 foam to exceed the friction resistance during the 0–15 mm displacement. Considering the 0–20 mm displacement period, the sum of the friction energy demand was 53.98 J. Tables 6 and 7 summarise the results of energy requirements.

Table 6. Energy requirements by 0.4 g/cm^3 foam given in Joules.

Stroke [mm]	Free Compr. Test	Radial Const. Test	Differential	Energy Requirement to Overcome the Friction
0–5	20.87 J	23.86 J	2.99 J	0.6 J
0–10	43.48 J	53.37 J	9.89 J	1.975 J
0–15	68.88 J	94.42 J	25.54 J	4.51 J
0–20	97.51 J	149.79 J	52.28 J	8.52 J

Table 7. Energy requirements by 0.7 g/cm^3 foam given in Joules.

Stroke [mm]	Free Compr. Test	Radial Const. Test	Differential	Energy Requirement to Overcome the Friction
0–5	31.53 J	59.84 J	28.31 J	1.08 J
0–10	72.21 J	142.93 J	70.72 J	9.86 J
0–15	121.54 J	244.03 J	122.49 J	28.22 J
0–20	190.03 J	415.18 J	225.03 J	53.98 J

10. Discussion

The compression test is a useful investigation form to reveal the mechanical behaviour of aluminium foam under loading, such as plateau stress, energy absorbing capacity and energy absorbing efficiencies, which are crucial evaluation considerations during crashworthiness development. Many studies are about free compression tests of metal foams, but the number of investigations about radial-constrained loads is moderate. The foam compression under radial constrain demands significantly more energy. This energy surplus is caused by the friction between the foam and the wall of constraint and the inhibition of radial deformation of the specimen. However, in the literature, there is no information about this friction resistance during the radially constrained test; therefore, the main goal of the present study was to reveal this friction resistance. Considering the results of the present investigations, the next conclusion can be stated.

Using the CFE (Crushing Force Efficiency) and CFF (Crushing Force Fluctuation) as the evaluation criteria to analyse the mechanical and folding behaviour of foam during radial-constrained loading is not relevant, since they show more than 100% due to the steepness of the end of the plateau zone; therefore, they can result in misunderstanding and deceptive results during the analysis. Instead of these conditions, the analysis must focus on the energy levels and strain of densifications. The CFE and CFF relevant values are collected in Table 2.

In this study, to obtain a relevant comparison between free and radial-constrained tests, the ISO13314:2011 standard was applied and taken into consideration. The standard characterises the method of calculation of plateau stress, plateau zone and the starting strain of densification concerning the free compression test, but there is no determination of conditions of the radial-constrained load, materials and friction parameters. Therefore, further friction tests were necessary to obtain a meaningful answer on the effect of radial forcing.

The foam with a density of 0.4 g/cm^3 absorbed an average of 31.2% more energy during the radially constrained compressive test compared to the free compressive one. The plateau range is a particularly important period of foam compression investigation; therefore, the difference in absorbed energy was also determined until the end of the plateau zone. In this case, the radially blocked foam absorbed 93.96 J, while the radially inhibited case absorbed 104.33 J, which means an 11.1% increase. Based on these, it can be stated that the significant energy surplus occurs in the densification zone and not in the plateau.

The test ranges were divided into different ranges by 5 mm displacements, such as 0–5 mm; 0–10 mm; 0–15 mm; and 0–20 mm. The total absorbed energy and the energy due to friction resistance are accounted for in all ranges. Based on the values via friction tests, the initial hypothesis has been confirmed, according to which the Coulomb friction during the compression inside a tube is influenced by the strain of the foam, the size of the contact surface between foam and constrain, and the Poisson ratio of the foam. Considering the values of Table 6, it can be stated that 18% of the invested energy was used to overcome friction in the case of the radial-constrained foam compression test.

The free compression test of 0.7 g/cm^3 foam required 126.53 J up to the end of the plateau zone, which is less by 44% compared to the radial-constrained version. Numerically, the constrained version demanded 225.31 J of energy before the beginning of the densification zone. The energy difference is 98.78 J, 39.64% of which was used exclusively to overcome the friction between the tube wall and the foam.

The measurement proved that the foam with a higher density can absorb more energy per unit displacement, but taking into account the SEA value introduced due to mass optimisation, it can be seen that the foam with a density of 0.4 g/cm^3 proves to be more efficient. Therefore, it is important to express how much energy a unit mass of foam can absorb per unit displacement. Based on these, it can be stated that if the goal in the design of the folding structure is to achieve a higher absorption capacity, then it is not the goal to use metal foam with a higher density in the structures. To reveal more information about the friction between the foam and tube, a surface roughness investigation is suggested to consider the wall of the tube before and after the test. Furthermore, an inner-lubricated version could provide useful information about the friction resistance proportion during the radial-constrained foam test.

11. Conclusions

The present work investigates the friction phenomena of a closed-cell aluminium foam-filled Al tube through compression testing. Between 0.43 and 0.7 g/cm^3 density foam was used in the same test conditions. This research could be filling in a gap in the field of foam-filled crash box structures science, since it reveals the magnitude of friction during a radial-constrained metal foam compression test. The novelty of the present research is not only in the values, but also in the applied method used during the investigation.

The radial-constrained compression test of 0.43 g/cm^3 foam calls for more energy with, 53.6% more than the free compression type. An average of 16% of this extra energy is due to the friction between the pipe wall and the foam. Considering the results of the 0.7 g/cm^3 foam, when the foam was compressed inside the aluminium tube, the process required about two times more compression energy than the free compression version. However, 23% of this energy surplus occurred owing to the friction phenomena.

Under compression, increasing the density of aluminium foam leads to a shorter plateau zone. The mechanical characteristics of aluminium foam can be greatly enhanced by adding a desirable gradation in pore frequency and altering the pore distribution. For foams with a relative density of 0.7, this can increase the plateau stress by nearly 100% and improve energy absorption as well. Since aluminium foam's relative density and compressive strength and Young's modulus are related, aluminium foam's density plays a critical role in determining how it behaves mechanically. A higher density of aluminium foam typically translates into increased stiffness, strength, and capacity to absorb energy. However, according to the results of the present study, the authors state that it is more appropriate to choose foam with a lower density than 0.7 g/cm^3 for radially inhibited compaction, as the foam quickly reaches the densification zone during compaction, especially in the case of the radially inhibited version, and thus its distorting and misleading results have an effect on energy and friction tests. However, it was a useful comparison, since it was observed that in a case of higher relative density, the presence of friction phenomena during the radial compression test is more intensive.

Author Contributions: Conceptualisation, J.K. and T.A.K.; methodology, J.K.; software, J.K.; validation, J.K., formal analysis, T.A.K.; investigation, J.K.; resources, T.A.K.; data curation, J.K.; writing—original draft preparation, J.K.; writing—review and editing, J.K.; visualisation, J.K.; supervision, T.A.K.; project administration, T.A.K.; funding acquisition, J.K. All authors have read and agreed to the published version of the manuscript.

Funding: The research is supported by the ÚNKP-23-3 New National Excellence Program of the Ministry for Culture and Innovation from the National Research, Development and Innovation Fund.

Institutional Review Board Statement: Not applicable.

Informed Consent Statement: Not applicable.

Data Availability Statement: The original contributions presented in the study are included in the article, further inquiries can be directed to the corresponding author.

Conflicts of Interest: The authors declare no conflict of interest.

References

1. Komara, A.I.; Budiwantoro, B.; Setiawan, R. Cellular structure design and manufacturability for electric vehicle: A review. *Int. J. Sustain. Transp. Technol.* **2022**, *5*, 70–79. [CrossRef]
2. Gokhale, A.A.; Kumar, N.V.; Sudhakar, B.; Sahu, S.N.; Basumatary, H.; Dhara, S. Cellular Metals and Ceramics for Defence Applications. *Def. Sci. J.* **2011**, *61*, 567–575. [CrossRef]
3. Munusamy, R.; Barton, D.C. Lightweight impact crash attenuators for a small formula SAE race car. *Int. J. Crashworthiness* **2010**, *15*, 223–234. [CrossRef]
4. Cho, H.; Cho, J. Damage and penetration behaviour of aluminum foam at various impacts. *J. Cent. South Univ.* **2014**, *21*, 3442–3448. [CrossRef]
5. Madgule, M.; Sreenivasa, C.G.; Borgaonkar, A.V. Aluminium metal foam production methods, properties and applications—A review. *Mater. Today Proc.* **2023**, *77*, 673–679. [CrossRef]
6. Karuppasamy, R.; Barik, D. Production methods of aluminium foam: A brief review. *Mater. Today Proc.* **2021**, *37*, 1584–1587. [CrossRef]
7. Schäffler, P. Aluminium foam applications. In *Metal Foams: Fundamentals and Applications*; DEStech Publications, Inc.: Lancaster, PN, USA, 2013; pp. 363–379.
8. Fuganti, A.; Lorenzi, L.; Grønsund, A.; Langseth, M. Aluminum foam for automotive applications. *Adv. Eng. Mater.* **2000**, *2*, 200–204. [CrossRef]
9. Han, M.S.; Min, B.S.; Cho, J.U. Fracture properties of aluminum foam crash box. *Int. J. Automot. Technol.* **2014**, *15*, 945–951. [CrossRef]
10. Sharma, S.S.; Yadav, S.; Joshi, A.; Goyal, A.; Khatri, R. Application of metallic foam in vehicle structure: A review. *Mater. Today Proc.* **2022**, *63*, 347–353. [CrossRef]
11. Srinath, G.; Vadiraj, A.; Balachandran, G.; Sahu, S.N.; Gokhale, A.A. Characteristics of aluminium metal foam for automotive applications. *Trans. Indian Inst. Met.* **2010**, *63*, 765–772. [CrossRef]
12. Li, Y.G.; Wei, Y.H.; Hou, L.F.; Guo, C.L.; Yang, S.Q. Fabrication and compressive behaviour of an aluminium foam composite. *J. Alloys Compd.* **2015**, *649*, 76–81. [CrossRef]
13. Naeem, M.A.; Gábora, A.; Mankovits, T. Influence of the manufacturing parameters on the compressive properties of closed cell aluminum foams. *Period. Polytech. Mech. Eng.* **2020**, *64*, 172–178. [CrossRef]
14. Mankovits, T.; Varga, T.A.; Manó, S.; Kocsis, I. Compressive response determination of closed-cell aluminium foam and linear-elastic finite element simulation of μCT-based directly reconstructed geometrical models. *Strojniški Vestnik-J. Mech. Eng.* **2018**, *64*, 105–113.
15. Sharma, V.; Grujovic, N.; Zivic, F.; Slavkovic, V. Influence of porosity on the mechanical behaviour during uniaxial compressive testing on voronoi-based open-cell aluminium foam. *Materials* **2019**, *12*, 1041. [CrossRef]
16. Surace, R.; De Filippis, L.A. Investigation and comparison of aluminium foams manufactured by different techniques. In *Advanced Knowledge Application in Practice*; IntechOpen: London, UK, 2010; pp. 95–118.
17. Baumeister, J.; Banhart, J.; Weber, M. Aluminium foams for transport industry. *Mater. Des.* **1997**, *18*, 217–220. [CrossRef]
18. File, M.; Cseke, R.; Huri, D.; Balogh, G.; Mankovits, T. Finite element analysis of closed-cell aluminum foam approximated with Weaire-Phelan unit cell structure. *IOP Conf. Ser. Mater. Sci. Eng.* **2022**, *1246*, 012002. [CrossRef]
19. Zhylgeldiyev, A.; Chernyshov, D.; Haider, S.; Mankovits, T. Modelling of closed-cell aluminum foam using Weaire–Phelan unit cells. *Int. Rev. Appl. Sci. Eng.* **2024**. [CrossRef]

20. Zhang, X.; Tang, L.; Liu, Z.; Jiang, Z.; Liu, Y.; Wu, Y. Yield properties of closed-cell aluminum foam under triaxial loadings by a 3D Voronoi model. *Mech. Mater.* **2017**, *104*, 73–84. [CrossRef]
21. Nammi, S.K.; Myler, P.; Edwards, G. Finite element analysis of closed-cell aluminium foam under quasi-static loading. *Mater. Des.* **2010**, *31*, 712–722. [CrossRef]
22. Su, M.; Wang, H.; Hao, H. Axial and radial compressive properties of alumina-aluminum matrix syntactic foam filled thin-walled tubes. *Compos. Struct.* **2019**, *226*, 111197. [CrossRef]
23. Gioux, G.; McCormack, T.M.; Gibson, L.J. Failure of aluminum foams under multiaxial loads. *Int. J. Mech. Sci.* **2000**, *42*, 1097–1117. [CrossRef]
24. Peroni, L.; Avalle, M.; Peroni, M. The mechanical behaviour of aluminium foam structures in different loading conditions. *Int. J. Impact Eng.* **2008**, *35*, 644–658. [CrossRef]
25. Chen, W. Optimisation for minimum weight of foam-filled tubes under large twisting rotation. *Int. J. Crashworthiness* **2001**, *6*, 223–242. [CrossRef]
26. Nariman-Zadeh, N.; Darvizeh, A.; Jamali, A. Pareto optimization of energy absorption of square aluminium columns using multi-objective genetic algorithms. *Proc. Inst. Mech. Eng. Part B J. Eng. Manuf.* **2006**, *220*, 213–224. [CrossRef]
27. D'Alessandro, V.; Petrone, G.; De Rosa, S.; Franco, F. Modelling of aluminium foam sandwich panels. *Smart Struct. Syst.* **2014**, *13*, 615–636. [CrossRef]
28. Banhart, J.; Seeliger, H.W. Aluminium foam sandwich panels: Manufacture, metallurgy and applications. *Adv. Eng. Mater.* **2008**, *10*, 793–802. [CrossRef]
29. Alia, R.A.; Cantwell, W.J.; Langdon, G.S.; Yuen SC, K.; Nurick, G.N. The energy-absorbing characteristics of composite tube-reinforced foam structures. *Compos. Part B Eng.* **2014**, *61*, 127–135. [CrossRef]
30. Nammi, S.K.; Edwards, G.; Shirvani, H. Effect of cell-size on the energy absorption features of closed-cell aluminium foams. *Acta Astronaut.* **2016**, *128*, 243–250. [CrossRef]
31. Razboršek, B.; Gotlih, J.; Karner, T.; Ficko, M.; Razboršek, B.; Ficko JG TK, M. The influence of machining parameters on the surface porosity of a closed-cell aluminium foam. *Stroj. Vestn.-J. Mech. Eng* **2019**, *66*, 29–37. [CrossRef]
32. Matsumoto, R.; Tsuruoka, H.; Otsu, M.; Utsunomiya, H. Fabrication of skin layer on aluminum foam surface by friction stir incremental forming and its mechanical properties. *J. Mater. Process. Technol.* **2015**, *218*, 23–31. [CrossRef]
33. Palic, N.; Sharma, V.; Zivic, F.; Mitrovic, S.; Grujovic, N. Tribology study of aluminum-based foam. *J. Prod. Eng.* **2019**, *22*, 1–4. [CrossRef]
34. Wang, Y.; Ou, B.; Zhu, P.; Niu, B.; Guo, Y.; Zhi, Q. High mechanical strength aluminum foam epoxy resin composite material with superhydrophobic, anticorrosive and wear-resistant surface. *Surf. Interfaces* **2022**, *29*, 101747. [CrossRef]
35. BeiHai Composite. Available online: https://www.beihaicomposite.com/aluminum-castings/ (accessed on 9 May 2024).
36. Kováčik, J.; Marsavina, L.; Linul, E. Poisson's ratio of closed-cell aluminium foams. *Materials* **2018**, *11*, 1904. [CrossRef]
37. Li, C.; Fu, Y.; Wang, B.; Zhang, W.; Bai, Y.; Zhang, L.; Qi, L. Effect of pore structure on mechanical and tribological properties of paper-based friction materials. *Tribol. Int.* **2020**, *148*, 106307. [CrossRef]
38. Hintikka, J.; Lehtovaara, A.; Mäntylä, A. Non-Coulomb friction in gross sliding fretting conditions with aluminium bronze against quenched and tempered steel. *Tribol. Int.* **2014**, *79*, 151–161. [CrossRef]
39. Wang, C.; Ma, R.; Zhao, J.; Zhao, J. Calculation method and experimental study of coulomb friction coefficient in sheet metal forming. *J. Manuf. Process.* **2017**, *27*, 126–137. [CrossRef]
40. Fras, T.; Rusinek, A.; Pęcherski, R.B.; Bernier, R.; Jankowiak, T. Analysis of friction influence on material deformation under biaxial compression state. *Tribol. Int.* **2014**, *80*, 14–24. [CrossRef]
41. Müser, M.H.; Urbakh, M.; Robbins, M.O. Statistical mechanics of static and low-velocity kinetic friction. *Adv. Chem. Phys.* **2003**, *126*, 187–272.
42. Smart, F.M. Kinetic friction. *Am. J. Orthod. Dentofac. Orthop.* **2004**, *125*, A17. [CrossRef]
43. Benabdallah, H.S. Static friction coefficient of some plastics against steel and aluminum under different contact conditions. *Tribol. Int.* **2007**, *40*, 64–73. [CrossRef]
44. Grilec, K.; Maric, G.; Jakovljevic, S. A study on energy absorption of aluminium foam. *BHM Berg-Und Hüttenmännische Monatshefte* **2010**, *155*, 231–234. [CrossRef]
45. Li, J.; Ma, G.; Zhou, H.; Du, X. Energy absorption analysis of density graded aluminium foam. *Int. J. Prot. Struct.* **2011**, *2*, 333–349. [CrossRef]
46. Rajak, D.K.; Kumaraswamidhas, L.A.; Das, S. An energy absorption behaviour of foam filled structures. *Procedia Mater. Sci.* **2014**, *5*, 164–172. [CrossRef]
47. Li, Z.; Xi, C.; Jing, L.; Wang, Z.; Zhao, L. Effect of loading rate on the compressive properties of open-cell metal foams. *Mater. Sci. Eng. A* **2014**, *592*, 221–229. [CrossRef]
48. Jung, A.; Diebels, S. Modelling of metal foams by a modified elastic law. *Mech. Mater.* **2016**, *101*, 61–70. [CrossRef]
49. Ramamurty, U.; Paul, A. Variability in mechanical properties of a metal foam. *Acta Mater.* **2004**, *52*, 869–876. [CrossRef]
50. Byakova, O.V.; Vlasov, A.O.; Semenov, M.V.; Zatsarna, O.V.; Gnyloskurenko, S.V. Mechanical Behaviour of the Porous and Foam Aluminium in Conditions of Compression: Determination of Key Mechanical Characteristics. *Met. Phys. Latest Technol.* **2017**. [CrossRef]

51. Umer, R.; Balawi, S.; Raja, P.; Cantwell, W.J. The energy-absorbing characteristics of polymer foams reinforced with bamboo tubes. *J. Sandw. Struct. Mater.* **2014**, *16*, 108–122. [CrossRef]
52. Alia, R.A.; Guan, Z.W.; Haldar, A.K.; Cantwell, W.J. A numerical study of the energy-absorption characteristics of metal tube-reinforced polymer foams. *J. Sandw. Struct. Mater.* **2016**, *18*, 597–623. [CrossRef]
53. Harishkumar, S.; Sriraman, A. Design Optimization on Front Side Rail to Improve the Crashworthiness. *IOSR J. Mech. Civ. Eng.* **2020**, *17*, 21–27.
54. Prabhaharan, S.A.; Balaji, G.; Annamalai, K. Numerical simulation of crashworthiness parameters for design optimization of an automotive crash-box. *Int. J. Simul. Multidiscip. Des. Optim.* **2022**, *13*, 3. [CrossRef]
55. Dirgantara, T.; Gunawan, L.; Putra, I.S.; Sitompul, S.A.; Jusuf, A. Numerical and experimental impact analysis of square crash box structure with holes. *Appl. Mech. Mater.* **2013**, *393*, 447–452. [CrossRef]
56. San Ha, N.; Lu, G. Thin-walled corrugated structures: A review of crashworthiness designs and energy absorption characteristics. *Thin-Walled Struct.* **2020**, *157*, 106995.
57. Peng, Y.; Li, T.; Bao, C.; Zhang, J.; Xie, G.; Zhang, H. Performance analysis and multi-objective optimization of bionic dendritic furcal energy-absorbing structures for trains. *Int. J. Mech. Sci.* **2023**, *246*, 108145. [CrossRef]
58. Wu, J.; Zhang, Y.; Li, K.; Su, L. Origami-inspired metamaterials hierarchical structure with tailorable crushing behaviour. *Constr. Build. Mater.* **2022**, *345*, 128328. [CrossRef]
59. *ISO 13314:2011*; Mechanical Testing of Metals—Ductility Testing—Compression Test for Porous and Cellular Metals. International Standard: Geneva, Switzerland, 2021. Available online: https://www.iso.org/obp/ui/#iso:std:iso:13314:ed-1:v1:en (accessed on 9 May 2024).
60. Yu, Y.; Cao, Z.; Tu, G.; Mu, Y. Energy absorption of different cell structures for closed-cell foam-filled tubes subject to uniaxial compression. *Metals* **2020**, *10*, 1579. [CrossRef]

materials

MDPI

Article

Homogeneous Age-hardening of Large-sized Al-Sc Foams via Micro-alloying with Zr and Ti

Xuming Chu [1,2,†], Tianze Wang [1,†], Donghui Yang [3], Xiangyang Peng [4], Shuo Hou [4], Shuai Chen [4], Guangyao Lu [4], Meiyuan Jiao [1], Yuan Wu [1,*], Andrey A. Rempel [5], Wentao Qu [6,*], Hongxiang Li [1] and Hui Wang [1,*]

1 State Key Laboratory for Advanced Metals and Materials, University of Science and Technology Beijing, Beijing 100083, China
2 Beijing Hengxing Yikang Technology Co., Ltd., Beijing 100191, China
3 College of Mechanics and Materials, Hohai University, Nanjing 210098, China
4 Equipment Research Center, China Nuclear Powder Technology Research Institute Co., Ltd., Shenzhen 518000, China
5 Institute of Metallurgy of the Ural Branch of the Russian Academy of Sciences, 620016 Ekaterinburg, Russia
6 School of Mechanical Engineering, Xi'an Shiyou University, Xi'an 710065, China
* Correspondence: wuyuan@ustb.edu.cn (Y.W.); wtqu@xsyu.edu.cn (W.Q.); wanghui@ustb.edu.cn (H.W.)
† These authors contributed equally in this work.

Abstract: Al-based foams have drawn increasing attention from industry due to their integration of structure and functional properties. However, large-sized Al-based foams still cannot be homogeneously strengthened by long-time aging due to their low thermal conductivity. In this study, we proposed an age-hardening approach that was applied in large-sized Al-0.16Sc-0.17Zr (wt.%) foams via micro-alloying with Zr and Ti compared with Al-0.21Sc foams; it not only achieved homogeneous strength by long-term aging but also reduced the cost of the alloy by substituting Zr and Ti for the more expensive Sc content. The results show that the Al_3(Sc, Zr, Ti) phase with a core–shell structure as a crucial precipitation strengthening phase by micro-alloying with Zr and Ti was less prone to coarsening after a prolonged aging heat treatment. Therefore, the yielding strength of Al-Sc foam micro-alloying with Zr and Ti remained almost unchanged after a maximum aging time of 1440 h due to less coarsening precipitate, which is consistent with the results of mechanical experiments. These findings provide a new way for the heat treatment strengthening of large-sized Al-based foams, thus promoting their industrial applications.

Keywords: Al-Sc foams; precipitates; age-hardening; micro-alloying; large size

check for updates

Citation: Chu, X.; Wang, T.; Yang, D.; Peng, X.; Hou, S.; Chen, S.; Lu, G.; Jiao, M.; Wu, Y.; Rempel, A.A.; et al. Homogeneous Age-hardening of Large-sized Al-Sc Foams via Micro-alloying with Zr and Ti. *Materials* **2024**, *17*, 1269. https://doi.org/10.3390/ma17061269

Academic Editor: Jordi Sort

Received: 26 January 2024
Revised: 6 March 2024
Accepted: 7 March 2024
Published: 9 March 2024

1. Introduction

As a class of ultra-lightweight materials, Al-based foams consisting of solid Al and containing a large volume fraction of gas-filled sealed pores have great potential to be utilized in industrial fields, such as automobile, aerospace, shipment, railway and civil construction, due to their unusual properties [1–5], i.e., good energy absorption capacity; excellent damping insulation of vibration, sound and heat; and low density. Many approaches can be employed to manufacture Al-based foams. Among which, only two methods, namely, melt foaming and powder metallurgy, are industrialized as cost-effective. Nonetheless, the relative low compressive strength of the Al-based foams fabricated by the above two methods, which is usually less than 15 MPa, impedes widespread applications [6,7].

Generally, the mechanical performance of Al-based foams can be improved through either optimizing pore-structures or strengthening the skeleton metal [8–11]. For instance, for a predetermined porosity, the compressive yielding strength of Al-based foam is increased by either decreasing the porosity, which may have an impact on the lightweight characteristic, or using a strong Al alloy as a starting metal. Therefore, it is the case that alloying followed by an appropriate heat treatment is one of competitive ways to enhance the mechanical properties of Al-based foams [5,12–14]. Previous results showed that additions of a small amount of Sc can lead to remarkable improvements in the mechanical properties [15–18], and such improvements are normally attributed to the precipitation hardening of the Al_3Sc phase formed during aging after solution heat treatments [19]. It was found that Sc additions have a major impact on strengthening, as they form a fine distribution of spherical $L1_2$ dispersoids. However, these effects on the mechanical performance of Al-Sc foams have not been investigated in detail yet. In particular, this improvement is still difficult to achieve in precipitation-strengthened Al-Sc foams with a large size due to their low thermal conductivity. In addition, considering the cost of the additional Sc element, it remains a challenge to obtain high strength from a thermal treatment of large-sized Al-Sc foams with less Sc contained.

There has been extensive work on the addition of Zr together with Sc to allow for the formation of a core–shell $L1_2$ structure with a core rich in Sc and a shell rich in Zr [20]. As a result, the addition of Zr together with Sc improves the effectiveness of Sc in Al alloys as an inhibitor of recrystallization and increases the stability of the alloy during prolonged annealing at high temperatures. Therefore, the addition of Zr not only reduces the susceptibility of the Al_3Sc precipitates to coarsening, finally leading to the high thermal stability of the precipitate, but also reduces the cost of the alloy by substituting Zr and Ti for the more expensive Sc content [21].

In this study, the fabrication of cellular Al-Sc foam samples via micro-alloying with a portion of Zr and Ti instead of Sc was attempted, and then heat treatments of these foams were conducted. The ultra-long-time aging hardening mechanism required for large-sized Al-Sc foams was explored experimentally as well.

2. Materials and Methods

A melt-foaming method initially based on the ALPORAS© route was applied to fabricate the cellular Al-0.16Sc-0.17Zr (wt.%) foam samples (Al alloy containing 0.16 wt.% Sc and 0.17 wt.% Zr) with a porosity of ~78% [22]. The specific preparation procedure was elaborated in detail previously [23,24], which can be summarized as follows. A predetermined quantity of Al-0.16Sc-0.17Zr alloy (~1 kg) was melted in a crucible at a fixed temperature and then 2.0 wt.% Ca was introduced into the molten alloy to increase the viscosity of the melt. Then, the thickened Al-Sc melt was foamed by adding a blowing agent, namely, TiH_2 powder (2.0 wt.%), at a predetermined foaming temperature (T). During the fabrication process, a stirring device was placed inside the melt to ensure a homogeneous distribution of the thickening agent and the blowing agent, and the final pore structures of the products were controlled by adjusting the foaming temperature in the range of 680–710 °C. Finally, the crucible was removed from the furnace, and immediately following, the foamed melt was entirely put into a water tank to solidify it; then, the Al-Sc foam samples were obtained [25]. Meanwhile, cellular Al-0.21Sc foams without Zr micro-addition and pure Al foams were also fabricated by the same route for comparison. The cell wall compositions of the final foams were determined by using an inductive coupled plasma–atomic emission spectrometer (ICP-AES, Pekin-Elmer, Waltham, MA, USA). The porosity (Pr) of the foam samples was calculated according to the following equation [26]:

$$Pr(\%) = [V - (M/\rho_s)]/V \times 100\% \quad (1)$$

where ρ_s is the density of the matrix (2.72 g cm^{-3}), and M and V are the mass and volume of the specimen, respectively.

According to the previous research results [27–29], the thermal conductivities of the Al-based foams are about 2 Wm^{-1} K^{-1}, which are much lower than that of the Al matrix. Consequently, the core temperature of a large-sized Al-based foam bulk does not reach the ambient temperature as quickly as that of Al matrix while it is heated, resulting in heterogeneous heating of the metallic framework. In these cases, two Inconel-encased Ni/CrNi thermocouples in this case were inserted into the center of the samples, which were Al-Sc foams with sizes of Φ 30 × 40 mm and Φ 100 × 135 mm, as presented in Figure 1.

The temperatures of two thermocouples in the cores, which reflected the core temperature changes during the heating process, were recorded.

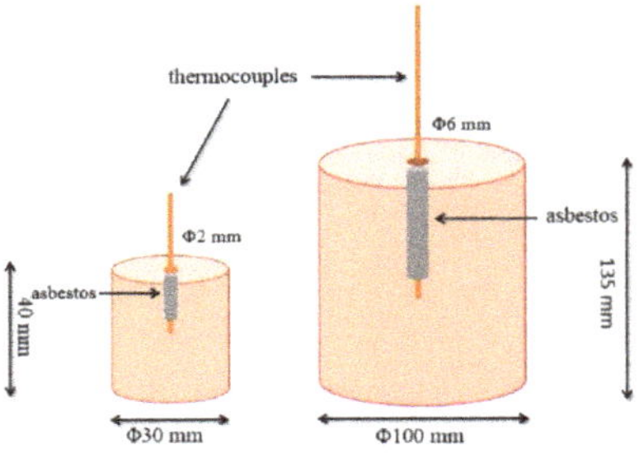

Figure 1. Schematic diagram of core temperature tests for Al-Sc foam samples with two sizes.

The microstructures of the cell walls of the foams were characterized by using an optical microscope (OM, ProgRes C5, Carl Zeiss SMT AG, Oberkochen, Germany), a scanning electronic microscope (SEM, SUPRA 55, Carl Zeiss SMT AG, Oberkochen, Germany) and a high-resolution transmission electronic microscope (HR-TEM, JEM 2100, JEOL Ltd., Tokyo, Japan).

The pore size of the fabricated foam was calculated by analyzing its scanned cross-sectional image with the aid of image analysis software. Briefly, the cross-sectional images of the foam were scanned by a scanner first. Since the intensity of reflected light from the cell edge is stronger than that from the inner surface of the pore, the obtained images can be easily analyzed using Image Pro Plus 3.0 software to acquire the plane pore areas. The equivalent diameter of a plane pore and the equivalent mean pore size can be subsequently calculated.

Specimens in dimension of Φ 24 × 35 mm for compressive tests were cut from the Al-based alloy foam with dimensions of Φ 30 × 40 by an electro-discharging machine, treated by isothermal aging heat treatments with a heating rate of 5 °C min^{-1} and kept from 200 to 600 °C for a period from 2 h to 1440 h (60 days). Some specimens with the same dimensions were also processed by isochronal aging from 300 to 500 °C for comparison. The quasi-static compressive experiments were carried out on the CMT 4350 Universal Testing Machine (Jinan Liangong Testing Technology Co., Ltd., Jinan, China) with a constant strain rate of 3 × 10^{-2} s^{-1} at a room temperature.

3. Results and Discussion

3.1. Pore Structure and Cell Wall Characterization of Fabricated Al-Sc Foams

As mentioned above, the mechanical properties of Al-based foams are influenced by the following factors: the pore structure, alloying additions and heat treatment. In these cases, the longitudinal sectional image of the foam sample is displayed in Figure 2a, which shows that it has a homogeneous pore structure distribution. The cross-sectional images of the Al-0.16Sc-0.17Zr foams with a porosity of ~78% are displayed in Figure 2b,c.Both the pore structure and porosity are distributed uniformly, indicating successful fabrication. Furthermore, for the Al-Sc alloys, the maximum solid solubility of Sc in Al was 0.38 wt.%, which occurred at the eutectic temperature, which is only 1 °C below the melting point of pure Al, and this low value is increased in practice to over 0.6 wt.% Sc due to non-equilibrium solidification conditions. When the concentration of Sc exceeds the solid solubility, the Al_3Sc primary phase forms during solidification and such a formation of this primary Al_3Sc phase can refine the grain size. As shown in Figure 3, the optical microscope images verified the grain sizes of cell walls of the foams, where Al-0.16Sc-0.17Zr and Al-0.21Sc were both approximately 80 μm, implying that the addition of 0.17%Zr had no obvious effect on the grain refinement compared with the Al-0.21Sc alloy without Zr addition.

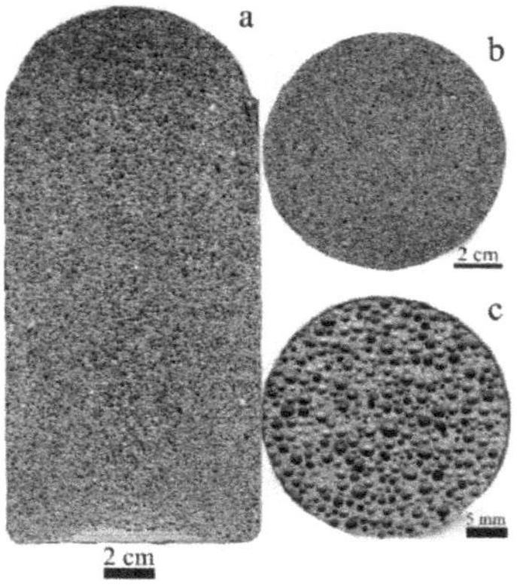

Figure 2. Section images of the cellular Al-0.16Sc-0.17Zr foam with the mean pore size of 1.5 mm: (**a**) longitudinal section image; (**b**) cross-sectional image; (**c**) cross-sectional image of the compressive sample with *Pr* = ~78%.

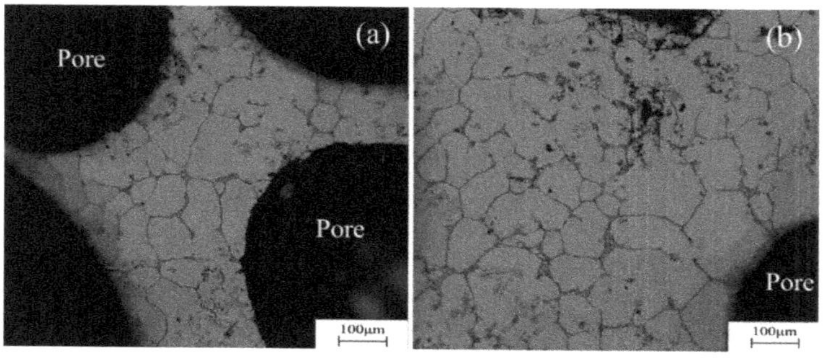

Figure 3. Microscopic structures of the cell walls: (**a**) Al-0.16Sc-0.17Zr foams; (**b**) Al-0.21Sc foams.

3.2. Inhomogeneity of Aging Temperature and Mechanical Properties of Al-Sc Foams

The temperature increases of the same Al-0.16Sc-0.17Zr foam samples with different sizes under the same heating conditions exhibited obvious differences in heat treatments, which are presented in Figure 4. The samples with the dimensions of Φ 100 × 135 mm and Φ 30 × 40 mm needed significantly different heating times to reach the temperature of 300 °C. Specifically, the Al alloy foams with the large size was heated to 300 °C after being heated for about 37 min from room temperature. In comparison, the sample with the small size only needed 16.5 min to reach the same temperature.

In these cases, which is known as Newtonian heating, the surface-area-to-volume ratio of the sample had a significant influence on the inhomogeneity in the temperature distribution. The temperature T in the sample can be expressed by

$$T(t) = T_{end} - (T_{end} - T_{start})\ exp(-t/t_{au}) \tag{2}$$

where t_{au} is proportional to the mass-to-surface ratio or, for a given porosity, to the volume-(V)-to-surface-(A) ratio. Herein, the proportionality constant is in essence the ratio of the heat transfer coefficient at the surface and the volumetric heat capacity, which, for a predetermined porosity (~78% in this case), is a constant. The surface-(A)-to-volume-(V) ratios (A/V) of the two samples with the dimensions of Φ 100 × 135 mm and Φ 30 × 40 mm were 0.55 cm^{-1} and 1.83 cm^{-1}, respectively, demonstrating that the heating of the small sample was about 3.3 times faster than that of the larger one. The results agreed with the different heating rates of the two samples in Figure 4, in which the heating rate of the small sample was 2.24 times faster than that of the large sample, considering other heating transferring effects. It can be concluded that the heating surface-area-to-volume ratio played an important role in the homogeneous temperature distribution rather than low thermal conductivity.

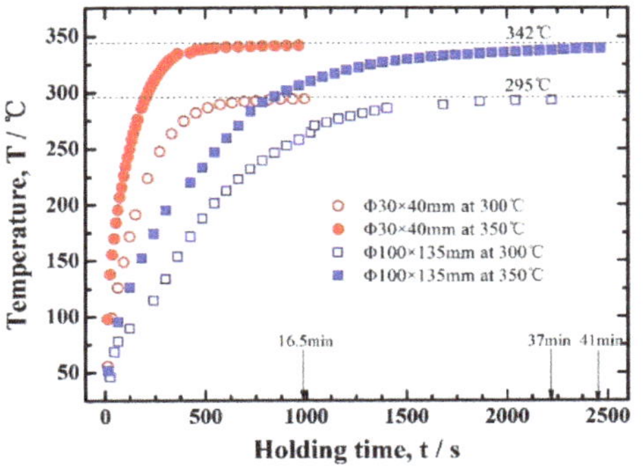

Figure 4. Core temperatures of the Al-0.16Sc-0.17Zr foams with different sizes during the heating at 300 and 350 °C.

In terms of an industrial large-sized Al foam bulk product, it usually has a dimensional size of 2.5 m × 1.5 m × 0.6 m, for which A/V was calculated as 0.055 cm^{-1}, which is much smaller than that of the above two samples. Therefore, homogeneous aging heat treatment needs a long time due to the low thermal conductivity rather than the A/V ratio.

To determine the appropriate peak aging temperature, the aging treatments on the two Al-Sc foams, namely, Al-0.16Sc-0.17Zr and Al-0.21Sc, in various temperatures were processed; then, quasi-static compression tests were conducted, where each test was repeated three times and the average of the numerical values was taken to ensure the accuracy. Compressive stress–strain curves of Al alloy foams, namely, the Al-0.16Sc-0.17Zr foams and Al-0.21Sc foams, with the same porosity of ~78% are shown in Figure 5a,b, respectively. These foams typically display three regions [30]: an initial, approximately linear deformation region until the peak stress; followed by a plateau region, where the stress was almost a constant; and the final densification region in which the stress increased steeply due to the fact that collapsed cells were almost fully compacted together. Herein, all 17 samples with different aging temperatures from 200 to 600 °C with steps of 25 °C were tested in Figure 5. The ends of the plateau range and final densification range were selected as the strains at 50% and 75%, respectively.

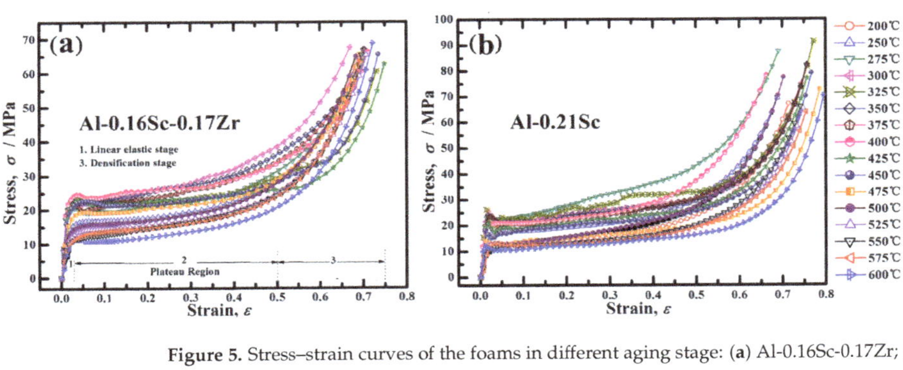

Figure 5. Stress–strain curves of the foams in different aging stage: (**a**) Al-0.16Sc-0.17Zr; (**b**) Al-0.21Sc.

3.3. Effect of Homogeneous Precipitates Distribution on the Yielding Strength of Al-Sc Foams Micro-Alloyed with Zr and Ti

An Al matrix can be precipitation-strengthened via micro-alloying with Sc, where the strength is improved by heat treatment aging [12]. Many studies have identified some substitutional elements, i.e., Zr and Ti, for Sc in the Al–Sc-based alloys to decrease the use of the expensive Sc, increase the ambient strength and increase the service temperature while maintaining a high strength [31–33]. Therefore, in general, the micro-addition of Zr is a highly attractive strategy in lowering the cost. In order to appropriately determine the heat treatment parameters for the Al-Sc-Zr/Al-Sc foams, differential scanning calorimetry (DSC, FDSC-800B, Shanghai Yanjin Scientific Instruments Co., Ltd., Shanghai, China) experiments for three samples, namely, as-cast Al-0.16Sc-0.17Zr, Al-0.21Sc and pure Al foams, were carried out in an Ar gas flow with 40 mL min^{-1} and at a heating rate of 5 °C min^{-1}. According to the Al-Ca phase diagram [34], α-Al and Al_4Ca undergo a eutectic reaction at 616 °C during the solidification process. As shown in Figure 6, for all Al-based foams, there was an endothermic peak starting at the temperature of 614 °C, which was most probably caused by the phase transformation of an intermetallic compound Al_4Ca, where the Ca was from dissolving the thickening agent and the transformation temperature of the Al_4Ca phase was approximately 616 °C. Furthermore, there was an additional endothermic peak that appeared around 634 °C for the Al-0.16Sc-0.17Zr foams [35], which was most probably caused by the phase transformation of primary Al_3(Sc, Zr, Ti) based on the research results of the Al-Sc-Zr-Ti alloys, although the phase transitions are not yet fully clear [36,37], where Ti was generated by the decomposition of the blowing agent, namely, TiH_2. With the temperature increase, all the alloy was melted, corresponding to the endothermic peaks with peak temperature of 660 °C due to the eutectic reaction $L \rightarrow \alpha\text{-}Al + Al_3Sc$. In this case, the solution treatment temperature should, therefore, be chosen to be below 614 °C in order to avoid over-burning because of the presence of Al_4Ca.

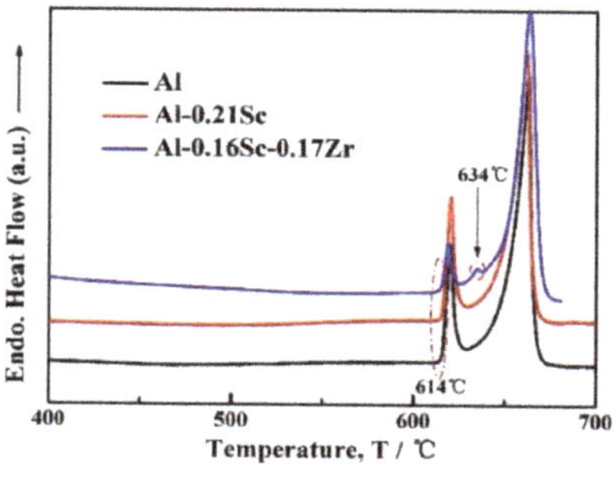

Figure 6. DSC curves of as-cast Al-0.16Sc-0.17Zr foams, as-cast Al-0.21Sc foams and Al foams.

TEM investigations for the precipitates in the cell walls of the Al-0.16Sc-0.17Zr foams after the isothermal aging at 400 °C for 3 h, 24 h and 120 h were conducted, as presented in Figure 7, and the Al-0.21Sc foams without Zr micro-alloying were processed similarly for comparison. As shown in Figure 7a–c, which were taken along the $[1\bar{1}1]$ zone axis, and in the selected area diffraction pattern (Figure 7g), the dark-field images of the Al-0.16Sc-0.17Zr foams indicate a clearly homogeneous distribution of the Al_3(Sc, Zr, Ti) precipitates, and such precipitates constantly kept a small size after the corresponding aging times for 3 h, 24 h and 120 h. In comparison, the dark-field images of the Al-0.21Sc foams had no Zr micro-alloying, as indicated in Figure 7d–f, which were taken from the $[011]$ zone axis (Figure 7h); this demonstrates that both precipitates were a little bit coarsened after aging times of 3 h, 24 h and 120 h.

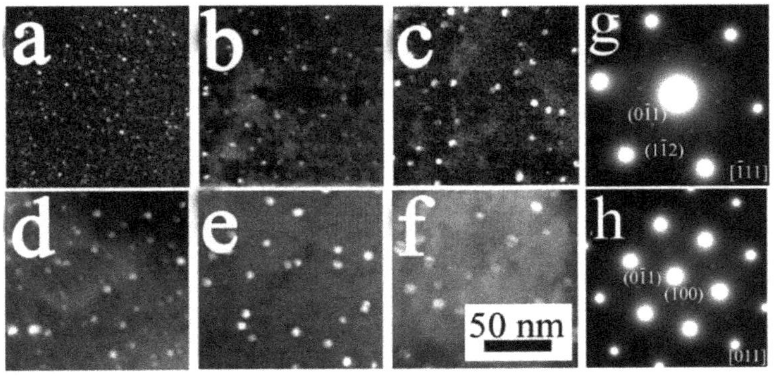

Figure 7. Comparison of precipitates, as observed by employing TEM images (utilizing the $0\bar{1}1$ superlattice reflection in the $[1\bar{1}1]$ and $[011]$ zone axes) of the Al-0.16Sc-0.17Zr (**a–c**) and Al-0.21Sc foams (**d–f**) after isothermal aging at 400 °C for 3 h, 24 h and 120 h, respectively; (**g**,**h**) are selected area electron diffraction images for Al-0.16Sc-0.17Zr foams and Al-0.21Sc foams, respectively. The magnification scale for the figures is the same as in (**f**).

The multi-faceted nature of Al_3Sc precipitates was reported in great detail [38,39], and $Al_3(Sc, Zr)$ precipitates with a cuboid core–shell structure in Al-Sc alloys with the micro-addition of Zr were also observed. During the aging of the Al-Sc micro-alloying with Zr, the Zr was pushed to the boundaries of the precipitate as the Al_3Sc precipitate continued to grow. The content of Zr around the precipitate was also enriched at a significant composition gradient, along with the diffusion of Zr in the matrix. The Zr in the precipitate and the matrix aggregated at the edge of the precipitate with the growth of the precipitate. This could form a stable core–shell structure with the Al_3Sc precipitate as the core and the $Al_3(Sc, Zr)$ precipitate as the shell. The presence of these core–shell cuboids can be counter-intuitive, as one would expect that the segregation of Zr would decrease the interfacial free energy, which would lead to the stabilization of the spheroidal particles. In fact, the core–shell structure was driven by the significant difference in diffusivities between the Sc and Zr. The use of multi-step aging treatments was found to be useful in separating the formation of the core at lower temperature and the shell at higher temperature to form a distribution of small and thermally stable precipitates. As shown in Figure 7, the precipitates at different aging times indicated that this core–shell structure of $Al_3(Sc, Zr, Ti)$ was more stable than that of Al_3Sc. The growth rate of the Al-Sc-Zr alloy was reduced to have better thermal stability performance.

Previous studies verified that Zr and Ti can also substitute for Sc in $L1_2$-ordered $Al_3(Sc, Zr, Ti)$ precipitates, and the chemical composition of Al_3Sc precipitates is anticipated to change with Zr additions due to Zr enrichment near the precipitate/matrix heterophase interface, which could alter the precipitate/dislocation interaction [40,41]. The concentration of the slow-diffuse Zr is sufficient to provide coarsening resistance at 400 °C for up to 66 days by forming a Zr-enriched outer shell that encapsulates the precipitates [42]. In this case, it can therefore be inferred that $Al_3(Sc, Zr)$ precipitates exhibited significantly higher coarsening resistance compared with Al_3Sc precipitates in the Al–Sc binary alloys owing to the sluggish diffusivity of Zr in α-Al [41]. A Zr-enriched precipitate shell was formed during the aging and served as a diffusion barrier for Sc, which was enriched in the precipitate. For microalloying with Ti, it could reduce the lattice parameter mismatch between the α-Al and Al_3Zr precipitates, which led to the corresponding interfacial energy decrease [33]. The diffusivity of Ti was lower than that of Zr and Sc, and therefore, it can be inferred that Ti could only replace some Zr in the outer shell of precipitates at a relative high aging temperature. In this case, as shown in Figure 8a, the core–shell structure precipitate, which is indicated by the red square, was observed in the cell wall of the Al-0.16Sc-0.17Zr foams after isochronal aging at 500 °C. Subsequently, it was supposed that the ordered-$L1_2$ precipitate formed such a core–shell structure, as shown in the schematic diagram in Figure 8b, was composed of a core of Al_3Sc enriched Sc, surrounded by a shell of Al_3(Sc, Zr, Ti) that was slightly enriched in Zr and Ti. The nanosized precipitation composition will be further characterized as a follow-up study.

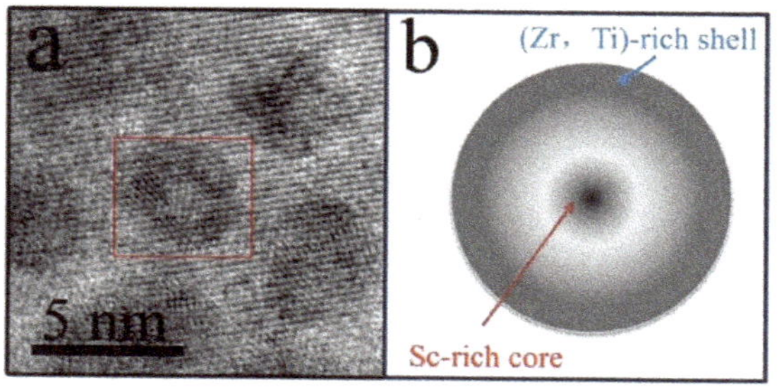

Figure 8. (**a**) HR-TEM image of Al_3(Sc, Zr, Ti) precipitates in Al-0.16Sc-0.17Zr after aging at 500 °C; (**b**) the supposed schematic diagram of Al_3(Sc, Zr, Ti) precipitate.

The size distribution of precipitates in the Al-0.16Sc-0.17Zr foam and Al-0.21Sc foam for comparison after direct isothermal aging at 400 °C for 3 h, 24 h and 120 h was measured to evaluate the coarsening of precipitates in Figure 9. Specifically, as a result, the precipitate radius of the Al-0.16Sc-0.17Zr foam at 400 °C for 3 h were distributed in the range of 0.5 to 2.5 nm, as displayed in Figure 9(a1), which, in comparison, was much smaller than that of the Al-0.21Sc foam in the range of 1.0 nm to 5.5 nm after the same heat treatment process, as displayed in Figure 9(b1). Moreover, the distributions of the precipitate radius in the Al-0.16Sc-0.17Zr foam for 24 h and 120 h, as shown in Figure 9(a2,a3), were 1.0 nm to 4.5 nm and 1.5 nm to 6.5 nm, respectively. In contrast, the distributions of the precipitate radius in the Al-0.21Sc foam for 24 h and 120 h, as displayed in Figure 9(b2,b3), were 3.0 nm to 7.5 nm and 4.0 nm to 10.5 nm, respectively. All the results also show that the Al_3(Sc, Zr, Ti) precipitates with this core–shell structure did not grow much after the prolonged aging and the micro-alloying with Zr and Ti compared with the Al-Sc foam with more Sc contents without Zr addition effectively restrained the kinetics of the precipitation coarsening.

The yielding strengths of two Al-Sc foams after different time spent aging at 300 °C are displayed in Figure 10, which also verified the effects of precipitation strengthening with long-term homogeneous aging on the mechanical properties. As shown in Figure 10, peak yield strengths of two foams, namely, Al-0.16Sc-0.17Zr foams and Al-0.21Sc foams, were achieved after 5 h of isothermal aging treatment. After the long-time aging processing, the yield strength of the Al-0.16Sc-0.17Zr foams remained almost consistent as the aging time increased, even after 1440 h, due the homogeneous precipitation strengthening effect. Meanwhile, the yielding strength of the Al-0.21Sc foams kept steadily decreasing compared with that of the Al-0.16Sc-0.17Zr foams, although the peak yielding strength value of the Al-0.21Sc foam (19.75 MPa) was higher than that of the Al-0.16Sc-0.17Zr foams (18.42 MPa) in the initial 24 h isothermal aging treatment. Notably, after the initial 24 h isothermal aging treatment, the yielding strength of the Al-0.21Sc foams was lower than that of the Al-0.16Sc-0.17Zr foams, while the yielding strength of the Al-0.16Sc-0.17Zr foams always maintained a high strength level as the aging time increased and finally surpassed the decreased strength of the Al-0.21Sc foams. The results also confirmed that Zr and Ti as substitutional elements for Sc in Al-Sc based alloys not only decreased the use of the expensive Sc, but also increased the ambient yielding strength of Al-Sc foams.

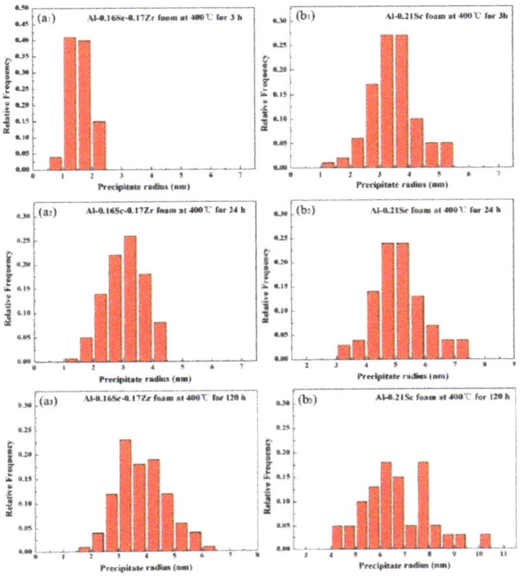

Figure 9. The size distributions of precipitates in (**a1–a3**) the Al-0.16Sc-0.17Zr foam and (**b1–b3**) the Al-0.21Sc foam after direct isothermal aging at 400 °C for 3 h, 24 h and 120 h.

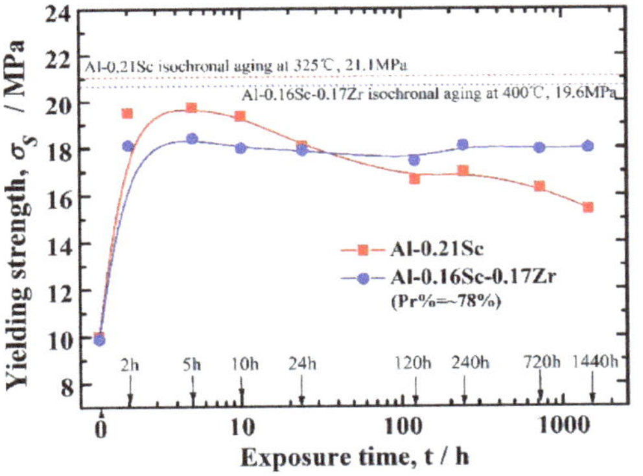

Figure 10. The yielding strengths of Al-Sc foams after different isothermal aging times at 300 °C.

Therefore, it was concluded that the yielding strength of Al-0.16Sc-0.17Zr foams could maintain unchanged level after a long time of aging, which is of special significance for large-sized Al-based foams to obtain homogeneous strength. In other words, the homogeneous precipitation of large-sized Al-Sc alloy foams was obtained via long-time aging heating, then these large-sized Al-Sc alloy foams were uniformly strengthened, which is more attractive for practical applications. In this case, the unchanged yielding strength of Al-0.16Sc-0.17Zr foams undergoing the isothermal aging at 300 °C within the period of 1440 h suggests that the large-sized Al alloy foams with micro-additions of Zr and Ti needed more aging time to be strengthened by the homogeneous precipitation heat treatment. In addition, the peak yielding strengths of Al-0.16Sc-0.17Sc foams and Al-0.21Sc foams after isochronal aging were 19.6 MPa and 21.1 MPa, respectively.

4. Conclusions

In this work, the effects of aging treatments on the microstructures and mechanical property of Al-0.16Sc-0.17Zr foams and Al-0.21Sc alloy foams with a porosity of ~78%, which were fabricated by a melt-foaming method, were investigated and we proposed an age-hardening approach, which was applied in large-sized Al-Sc alloy foams via micro-alloying with Zr and Ti. Owing to the Al_3(Sc, Zr, Ti) phase with a core–shell structure that was less prone to coarsening of homogeneous precipitates after prolonged aging, the yielding strength of Al-Sc foams micro-alloyed with Zr and Ti remained almost stable after a maximum of 1440 aging hours, which is of great significance for the heat treatment strengthening of large-sized Al-based foams. Our findings provide a new method for the heat treatment strengthening of large-sized Al-based foams, thus promoting their industrial applications.

Author Contributions: X.C. and T.W. designed the experiments. X.C. and D.Y. carried out the main experimental program, including the fabrication of the samples. X.P., S.H. and S.C. wrote the manuscript draft. T.W. and G.L. conducted the TEM and DSC characterizations, respectively. M.J. analyzed the TEM characterization data. A.A.R. and Y.W. reviewed the final manuscript. W.Q. proposed the concept and discussed the results. H.W. and H.L. reviewed and edited the manuscript and supervised the project. All authors reviewed and contributed to the final manuscript. All authors have read and agreed to the published version of the manuscript.

Funding: This research was financially supported by the Project of International Cooperation and Exchanges of the NSFC (grant no. 1251200014). WT Qu and H Wang acknowledge the financial support from the National Nature Science of Foundation of China (grant nos. 52071261 and 52271003) and the Key Research and Development Program of Shaanxi Province (grant no. 2022GY-401). Authors are very grateful to Gehong Deng from the Department of Foreign Language and Business, Guangdong Women's Polytechnic College for the language polishing.

Institutional Review Board Statement: Not applicable.

Informed Consent Statement: Not applicable.

Data Availability Statement: Data are contained within the article.

Conflicts of Interest: Author Xuming Chu was employed part-timely by the company Beijing Hengxing Yikang Technology Co., Ltd. Authors Xiangyang Peng, Shuo Hou, Shuai Chen and Guangyao Lu were employed by the company China Nuclear Powder Technology Research Institute Co., Ltd. The remaining authors declare that the research was conducted in the absence of any commercial or financial relationships that could be construed as a potential conflict of interest.

References

1. Wang, H.; Zhu, D.F.; Wu, Y.; Liu, X.J.; Jiang, S.H.; Nieh, T.G.; Lu, Z.P. New Insight into Fabrication of Shaped Mg–X Alloy Foams with Cellular Structure via a Gas Release Reaction Powder Metallurgy Route. *J. Iron Steel Res. Int.* **2021**, *28*, 125–132. [CrossRef]
2. Banhart, J.; Seeliger, H. Recent Trends in Aluminum Foam Sandwich Technology. *Adv. Eng. Mater.* **2012**, *14*, 1082–1087. [CrossRef]
3. García-Moreno, F. Commercial Applications of Metal Foams: Their Properties and Production. *Materials* **2016**, *9*, 85. [CrossRef]
4. Rao, D.W.; Yang, Y.W.; Huang, Y.; Sun, J.B.; Pan, L.W.; Hu, Z.L. Microstructure and Compressive Properties of Aluminum Matrix Syntactic Foams Containing Al_2O_3 Hollow Particles. *Met. Mater.* **2021**, *58*, 395–407. [CrossRef]
5. Byakova, A.; Gnyloskurenko, S.; Bezimyanniy, Y.; Nakamura, T. Closed-Cell Aluminum Foam of Improved Sound Absorption Ability: Manufacture and Properties. *Metals* **2014**, *4*, 445–454. [CrossRef]
6. Ferraris, S.; Santostefano, A.; Barbato, A.; Molina, R.; Ubertalli, G. Al-Based Foams as Permanent Cores in Al Castings: Effect of Surface Skin Thickness and Composition on Infiltration and Core-Shell Bonding. *Metals* **2021**, *11*, 1715. [CrossRef]
7. Sasikumar, S.; Georgy, K.; Mukherjee, M. Production, Stability, and Properties of In-situ Al-5ZrB_2 Composite Foams. *Mater. Sci. Eng. A* **2022**, *849*, 143501. [CrossRef]
8. Hangai, Y.; Ando, M.; Ohashi, M.; Amagai, K.; Suzuki, R.; Matsubara, M.; Yoshikawa, N. Compressive Properties of Two-Layered Aluminum Foams with Closed-Cell and Open-Cell Structures. *Mater. Today Commun.* **2020**, *24*, 101249. [CrossRef]
9. Almonti, D.; Baiocco, G.; Tagliaferri, V.; Ucciardello, N. Design and Mechanical Characterization of Voronoi Structures Manufactured by Indirect Additive Manufacturing. *Materials* **2020**, *13*, 1085. [CrossRef]
10. Byakova, A.V.; Gnyloskurenko, S.V.; Nakamura, T. Effect of $CaCO_3$ Foaming Agent at Formation and Stabilization of Al-Based Foams Fabricated by Powder Compact Technique. *Mater. Trans.* **2017**, *58*, 249–258. [CrossRef]

11. Sabuwala, T.; Gioia, G. Skeleton-and-Bubble Model of Polyether-Polyurethane Elastic Open-Cell Foams for Finite Element Analysis at Large Deformations. *J. Mech. Phys. Solids* **2013**, *61*, 886–911. [CrossRef]
12. Oh, J.-S.; Shim, M.-C.; Park, M.-H.; Lee, K.-A. High Temperature Oxidation Behavior of Ni-Cr-Al Based Powder Porous Metal. *Met. Mater. Int.* **2014**, *20*, 915–921. [CrossRef]
13. Pan, L.; Rao, D.; Yang, Y.; Qiu, J.; Sun, J.; Gupta, N.; Hu, Z. Gravity Casting of Aluminum-Al_2O_3 Hollow Sphere Syntactic Foams for Improved Compressive Properties. *J. Porous Mater.* **2020**, *27*, 1127–1137. [CrossRef]
14. Hangai, Y.; Saito, K.; Utsunomiya, T.; Kuwazuru, O.; Yoshikawa, N. Fabrication and Compression Properties of Functionally Graded Foam with Uniform Pore Structures Consisting of Dissimilar A1050 and A6061 Aluminum Alloys. *Mater. Sci. Eng. A* **2014**, *613*, 163–170. [CrossRef]
15. Davydov, V.G.; Rostova, T.D.; Zakharov, V.V.; Filatov, Y.A.; Yelagin, V.I. Scientific Principles of Making an Alloying Addition of Scandium to Aluminium Alloys. *Mater. Sci. Eng. A* **2000**, *280*, 30–36. [CrossRef]
16. Huang, L.; Wang, H.; Yang, D.; Ye, F.; Lu, Z.P. Effects of Scandium Additions on Mechanical Properties of Cellular Al-Based Foams. *Intermetallics* **2012**, *28*, 71–76. [CrossRef]
17. Gao, Y.H.; Cao, L.F.; Yang, C.; Zhang, J.Y.; Liu, G.; Sun, J. Co-Stabilization of θ'-Al_2Cu and Al_3Sc Precipitates in Sc-Microalloyed Al–Cu Alloy with Enhanced Creep Resistance. *Mater. Today Nano* **2019**, *6*, 100035. [CrossRef]
18. Hu, Z.; Wei, Z.; Hou, B.; Zhao, Y.; Pang, X.; Zhang, J.; Li, L. Effects of Sc Addition and Retrogression and Re-Aging on Microstructure and Mechanical Properties of Al-8Zn-2Mg-2Cu-0.15Zr Alloy. *J. Mater. Eng. Perform.* **2022**, *31*, 5409–5418. [CrossRef]
19. Huang, L.; Yang, D.H.; Wang, H.; Ye, F.; Lu, Z.P. Effects of Scandium on Corrosion Resistance and Mechanical Properties of Cellular Al-Based Foams. *Mater. Sci. Forum* **2013**, *747–748*, 93–100. [CrossRef]
20. Dorin, T.; Babaniaris, S.; Jiang, L.; Cassel, A.; Eggeman, A.; Robson, J. Precipitation Sequence in Al-Sc-Zr Alloys Revisited. *Materialia* **2022**, *26*, 101608. [CrossRef]
21. Tolley, A.; Radmilovic, V.; Dahmen, U. Segregation in Al_3(Sc,Zr) Precipitates in Al–Sc–Zr Alloys. *Scr. Mater.* **2005**, *52*, 621–625. [CrossRef]
22. Miyoshi, T.; Itoh, M.; Akiyama, S.; Kitahara, A. ALPORAS Aluminum Foam: Production Process, Properties, and Applications. *Adv. Eng. Mater.* **2000**, *2*, 179–183. [CrossRef]
23. Huang, L.; Wang, H.; Yang, D.H.; Ye, F.; Wang, S.Q.; Lu, Z.P. Effects of Calcium on Mechanical Properties of Cellular Al–Cu Foams. *Mater. Sci. Eng. A* **2014**, *618*, 471–478. [CrossRef]
24. Byakova, A.; Gnyloskurenko, S.; Vlasov, A.; Yevych, Y.; Semenov, N. The Mechanical Performance of Aluminum Foam Fabricated by Melt Processing with Different Foaming Agents: A Comparative Analysis. *Metals* **2022**, *12*, 1384. [CrossRef]
25. Yang, D.; Chen, J.; Wang, H.; Jiang, J.; Ma, A.; Lu, Z.P. Effect of Decomposition Kinetics of Titanium Hydride on the Al Alloy Melt Foaming Process. *J. Mater. Sci. Technol.* **2015**, *31*, 361–368. [CrossRef]
26. Qu, Y.-R.; Liu, S.-A.; Wu, H.; Li, M.-L.; Tian, H.-C. Tracing Carbonate Dissolution in Subducting Sediments by Zinc and Magnesium Isotopes. *Geochim. Cosmochim. Acta* **2022**, *319*, 56–72. [CrossRef]
27. Sadeghi, E.; Hsieh, S.; Bahrami, M. Thermal Conductivity and Contact Resistance of Metal Foams. *J. Phys. Appl. Phys.* **2011**, *44*, 125406. [CrossRef]
28. Wang, H.; Zhou, X.; Long, B.; Yang, J.; Liu, H. Thermal Properties of Closed-Cell Aluminum Foams Prepared by Melt Foaming Technology. *Trans. Nonferrous Met. Soc. China* **2016**, *26*, 3147–3153. [CrossRef]
29. Abishek, S.; King, A.J.C.; Nadim, N.; Mullins, B.J. Effect of Microstructure on Melting in Metal-Foam/Paraffin Composite Phase Change Materials. *Int. J. Heat Mass Transf.* **2018**, *127*, 135–144. [CrossRef]
30. Manoj; Afzal Khan, D.M.; Mondal, D.P. High Temperature Deformation Behavior of Closed Cell ZnAl27 Hybrid Foam Made through Stir Casting Technique. *Mater. Sci. Eng. A* **2018**, *731*, 324–330. [CrossRef]
31. Siddesh Kumar, N.M.; Dhruthi; Pramod, G.K.; Samrat, P.; Sadashiva, M. A Critical Review on Heat Treatment of Aluminium Alloys. *Mater. Today Proc.* **2022**, *58*, 71–79. [CrossRef]
32. Van Dalen, M.E.; Gyger, T.; Dunand, D.C.; Seidman, D.N. Effects of Yb and Zr Microalloying Additions on the Microstructure and Mechanical Properties of Dilute Al–Sc Alloys. *Acta Mater.* **2011**, *59*, 7615–7626. [CrossRef]
33. Van Dalen, M.E.; Dunand, D.C.; Seidman, D.N. Effects of Ti Additions on the Nanostructure and Creep Properties of Precipitation-Strengthened Al–Sc Alloys. *Acta Mater.* **2005**, *53*, 4225–4235. [CrossRef]
34. Binary Alloy Phase Diagrams. In *Alloy Phase Diagrams*; Okamoto, H.; Schlesinger, M.E.; Mueller, E.M., Eds.; ASM International: Detroit, MI, USA, 2016; Volume 3. [CrossRef]
35. Vlach, M.; Stulíková, I.; Smola, B.; Žaludová, N.; Černá, J. Phase Transformations in Isochronally Annealed Mould-Cast and Cold-Rolled Al–Sc–Zr-Based Alloy. *J. Alloys Compd.* **2010**, *492*, 143–148. [CrossRef]
36. Feng, J.; Ye, B.; Zuo, L.; Qi, R.; Wang, Q.; Jiang, H.; Huang, R.; Ding, W.; Yao, J.; Wang, C. Effects of Zr, Ti and Sc Additions on the Microstructure and Mechanical Properties of Al-0.4Cu-0.14Si-0.05Mg-0.2Fe Alloys. *J. Mater. Sci. Technol.* **2018**, *34*, 2316–2324. [CrossRef]
37. Bo, H.; Liu, L.B.; Hu, J.L.; Jin, Z.P. Experimental Study and Thermodynamic Modeling of the Al-Sc-Zr System. *Comput. Mater. Sci.* **2017**, *133*, 82–92. [CrossRef]
38. Watanabe, C.; Watanabe, D.; Monzen, R. Coarsening Behavior of Al_3Sc Precipitates in an Al-Mg-Sc Alloy. *Mater. Trans.* **2006**, *47*, 2285–2291. [CrossRef]

39. Na, X.; Wenqing, L.; Liu, Z.; Muthuramalingam, T. Effect of Scandium in Al–Sc and Al–Sc–Zr Alloys Under Precipitation Strengthening Mechanism at 350 °C Aging. *Met. Mater. Int.* **2021**, *27*, 5145–5153. [CrossRef]
40. Van Dalen, M.E.; Seidman, D.N.; Dunand, D.C. Creep- and Coarsening Properties of Al–0.06at.% Sc–0.06at.% Ti at 300–450 °C. *Acta Mater.* **2008**, *56*, 4369–4377. [CrossRef]
41. Fuller, C.B.; Seidman, D.N.; Dunand, D.C. Mechanical Properties of Al(Sc,Zr) Alloys at Ambient and Elevated Temperatures. *Acta Mater.* **2003**, *51*, 4803–4814. [CrossRef]
42. Vo, N.Q.; Dunand, D.C.; Seidman, D.N. Improving Aging and Creep Resistance in a Dilute Al–Sc Alloy by Microalloying with Si, Zr and Er. *Acta Mater.* **2014**, *63*, 73–85. [CrossRef]

materials

MDPI

Article

Aggregation–Growth and Densification Behavior of Titanium Particles in Molten Mg-$MgCl_2$ System

Xin Yang [1], Kaihua Li [1,2,4,*], Jun Li [1,*], Zhuo Sheng [2,3,4] and Ying Liu [1]

1 School of Materials Science and Engineering, Sichuan University, Chengdu 610065, China; yangxin15680824707@163.com (X.Y.)
2 Key Laboratory of Vanadium and Titanium Resources Comprehensive Utilization, Panzhihua 617000, China
3 Faculty of Metallurgical and Energy Engineering, Kunming University of Science and Technology, Kunming 650093, China
4 Pangang Group Research Institute Co., Ltd., Panzhihua 617000, China
* Correspondence: lkh002@163.com (K.L.); abthonyli@163.com (J.L.)

Abstract: In this work, the preparation of titanium sponge by magnesium thermal method is regarded as the liquid-phase sintering process of titanium, and powder-metallurgy sintering technology is utilized to simulate the aggregation–growth and densification behavior of titanium particles in a high-temperature liquid medium (the molten Mg-$MgCl_2$ system). It was found that compared with $MgCl_2$, Mg has better high-temperature wettability and reduction effect, which promotes titanium particles to form a sponge titanium skeleton at lower temperature. The aggregation degree of titanium particles and the densification degree of a sponge titanium skeleton can be improved by increasing the temperature and the relative content of Mg in the melting medium. The kinetics study shows that with the increase in temperature, the porosity of the titanium particle aggregates and the sponge titanium skeleton decreases, and their density growth rate increases. With the extension of time, the aggregation degree of titanium particles and the densification degree of sponge titanium gradually increase. This work provides a theoretical reference for controlling the density of titanium sponge in industry.

Keywords: titanium sponge; melting medium; aggregation and growth; pore structure; densification; kinetics

check for updates

Citation: Yang, X.; Li, K.; Li, J.; Sheng, Z.; Liu, Y. Aggregation–Growth and Densification Behavior of Titanium Particles in Molten Mg-$MgCl_2$ System. *Materials* **2024**, *17*, 2904. https://doi.org/10.3390/ma17122904

Academic Editor: Jan Haubrich

Received: 7 May 2024
Revised: 4 June 2024
Accepted: 10 June 2024
Published: 13 June 2024

1. Introduction

Titanium and titanium alloys have significant applications in aerospace, marine development, petrochemical industry, and clinical medical treatment. The efficient preparation of high-quality sponge titanium has become a concern [1]. In industrial production, the thermal reduction of titanium using Mg metal is an important method: under the protection of argon, titanium tetrachloride ($TiCl_4$) is reduced by Mg metal to produce titanium sponge, and excess Mg and residual $MgCl_2$ are then removed through vacuum distillation. The main reactions involved are as follows:

$$TiCl_4(g) + 2Mg(l) = 2MgCl_2(l) + Ti(s)$$

Akihiro Kishimoto and Tetsuya Uda [2] observed the in situ process from the addition of $TiCl_4$ to the reaction between Mg and $TiCl_4$, resulting in the formation of sponge titanium. They also studied the reaction mechanism of sponge titanium. Ch.R.V.S. Nagesh et al. [3] examined the stacking process of titanium particles in different stages of the magnesium thermal reduction reaction. Numerous reports have been published on the preparation of sponge titanium through the magnesium thermal reduction of $TiCl_4$, and researchers have proposed various reaction mechanisms [4–12]. However, in general, the production process involves the following steps: the magnesium-based thermal reaction leading to the precipitation of fine titanium particles, gradual growth of these particles, aggregation and growth of titanium particles at high temperatures to form porous sponge titanium, and gradual densification of the porous sponge titanium in the molten Mg-$MgCl_2$ system to form titanium sponge lumps [13–18]. However, the titanium sponge lumps produced using this method often exhibit uneven density distribution, particularly with a dense and hard core in the middle and lower parts. This results in a dense structure and coarse particles in titanium sponge, which can lead to composition segregation and structural defects in the preparation of titanium and titanium alloys. Consequently, controlling the particle size and pore structure of sponge titanium during production is a significant challenge in current research.

The process of producing sponge titanium through the magnesium thermal reduction reaction is highly complex, involving high temperatures, sealing, and kinetics imbalances [19,20]. Directly studying the laws governing the aggregation–growth and densification of titanium particles in industrial batch production is extremely difficult. In this work, we propose considering the process of titanium particle aggregation–growth to form sponge titanium as a liquid-phase sintering process of titanium particles in a molten Mg-$MgCl_2$ system. Powder-metallurgy sintering technology is utilized to simulate the aggregation–growth and densification behavior of titanium particles in a high-temperature liquid medium. Therefore, this paper investigates the influence of medium composition on the aggregation–growth and densification behavior of titanium particles in the molten Mg-$MgCl_2$ system. The findings provide a theoretical reference for controlling the particle size and pore structure of sponge titanium.

2. Materials and Methods

2.1. Materials

Pure Mg metal (purity $\geq$ 99.9%, provided by Qinghe Dingyuan Metal Products Co., Ltd., Xingtai, China), anhydrous $MgCl_2$ (purity $\geq$ 99.9%, provided by Shanghai Aladdin Biochemical Technology Co., Ltd., Shanghai, China), and micron-sized titanium powder (purity $\geq$ 99.9%, D_{50} = 28.1 μm, provided by Zhejiang Yamei Nano Technology Co., Ltd., Jiashan, China) were used as raw materials.

2.2. Sample Preparation

Titanium powder, pure magnesium, and anhydrous magnesium chloride were added to a corundum crucible and thoroughly mixed before being placed into a vertical atmosphere furnace (SK-5-17Q, provided by FNS (Beijing) Electric Furnace Co., Ltd, Beijing, China) for sintering. In industrial production of titanium sponge, the reaction vessel experiences a temperature range from 750 °C to 1200 °C, with higher temperatures at the reaction center area [21,22]. As Mg undergoes thermal reduction, the molten Mg gradually reacts, while molten $MgCl_2$ and sponge titanium are continuously generated. To simulate the aggregation–growth process of titanium particles, sintering temperatures were set at 800 °C, 900 °C, 1000 °C, 1100 °C, 1200 °C, and 1300 °C. The sintering media used were molten $MgCl_2$ media and molten $MgCl_2$:Mg (1:1) media. Before the sintering process, the equipment underwent three rounds of Ar gas purging. During the sintering process, Ar gas was continuously introduced at a flow rate of 1 L/min to prevent titanium particles from making contact with air and oxidizing at high temperatures. After sintering, the samples were subjected to vacuum distillation at 750 °C for 2 h to remove residual Mg and $MgCl_2$ and obtain pure sponge titanium.

2.3. Characterization

Samples collected from different parts of the titanium sponge after sintering–distillation were analyzed. The high-temperature contact angles of the Mg-Ti and $MgCl_2$-Ti systems were measured using a high-temperature contact angle measuring instrument (KRUSS DSAHT17C, Hamburg, Germany) to characterize the differences in wettability of different media on sponge titanium. The oxygen content of titanium particles and the composition of the oxide layer were characterized using a nitrogen–oxygen analyzer (ON-3000, Beijing, China) and X-ray photoelectron spectroscopy (XPS, K-Alpha, Waltham, MA, USA), respectively. The area percentage of titanium particles was measured using a metallographic microscope (Axio Observer A5m, Hanover, Germany) combined with Image-Pro Plus 6.0 software to assess the degree of aggregation of titanium particles. The average size change of titanium particle aggregates was measured using a laser particle size analyzer (Malvern Mastersizer 3000, Malvern, UK), and the growth behavior of aggregates in different media and at different temperatures was analyzed. The microstructure of the samples was observed using a field emission scanning electron microscope (ZEISS Sigma 500 FE-SEM, Oberkochen, Germany) to analyze the differences in the microstructure of titanium particles in different media. The total pore area, average pore size, and porosity of the samples were measured using a mercury intrusion meter (MicroActive AutoPore V 9600, Purchase, NY, USA) to analyze the changes in the pore structure of sponge titanium.

3. Results and Discussion

3.1. Influencing Factors in Liquid-Phase Sintering

Figure 1 presents the results of the Mg-Ti wettability test. As the temperature rises from room temperature to 850 °C, the Mg remains solid and its shape remains unchanged. When the temperature reaches 900 °C, the Mg gradually melts. After holding at 900 °C for 1 min, the molten Mg starts to spread on the surface of the titanium substrate, resulting in a significant decrease in the contact angle. The contact angle of the molten Mg on the Ti sheet continues to decrease and eventually stabilizes at around 3.1° when the holding time is extended. Figure 2 shows the test results of the $MgCl_2$-Ti wettability. As the temperature rises from room temperature to 714 °C, the shape of the $MgCl_2$ remains unchanged as it has not yet melted. At 900 °C, the $MgCl_2$ begins to melt, and its appearance undergoes a noticeable change. After holding at 900 °C for 1 min, the $MgCl_2$ completely melts and spreads on the surface of the titanium substrate, leading to a significant decrease in the contact angle. When the holding time is extended to 5 min, the molten $MgCl_2$ spreads completely on the titanium sheet, and the contact angle decreases to 6.9°. The contact angle of Mg-Ti is smaller than that of $MgCl_2$-Ti, indicating that molten Mg has better wettability with Ti at high temperatures compared to molten $MgCl_2$. This suggests that Ti atoms diffuse more easily in molten Mg [23,24]. On the other hand, better wettability of Mg/Ti means that molten Mg spreads more fully on the surface of titanium particles. The capillary force between adjacent titanium particles is greater, promoting the particles to come into contact with each other and form a sponge titanium framework, thereby accelerating the aggregation speed of titanium particles.

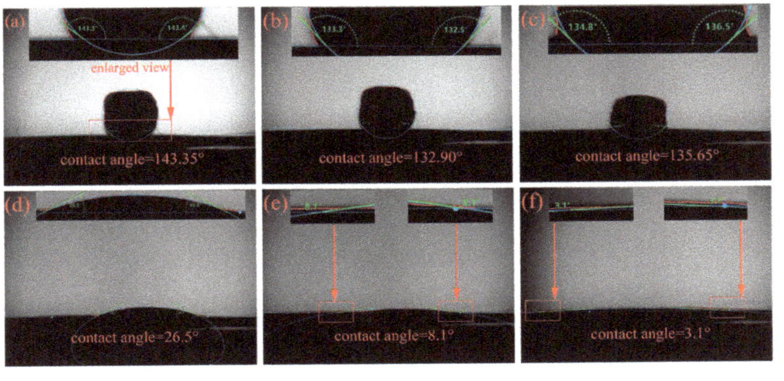

Figure 1. Photographs of the Mg-Ti contact angle test: (**a**) 25 °C, (**b**) 850 °C, (**c**) 900 °C, (**d**) holding the temperature at 900 °C for 1 min, (**e**) holding the temperature at 900 °C for 3 min, and (**f**) holding the temperature at 900 °C for 5 min.

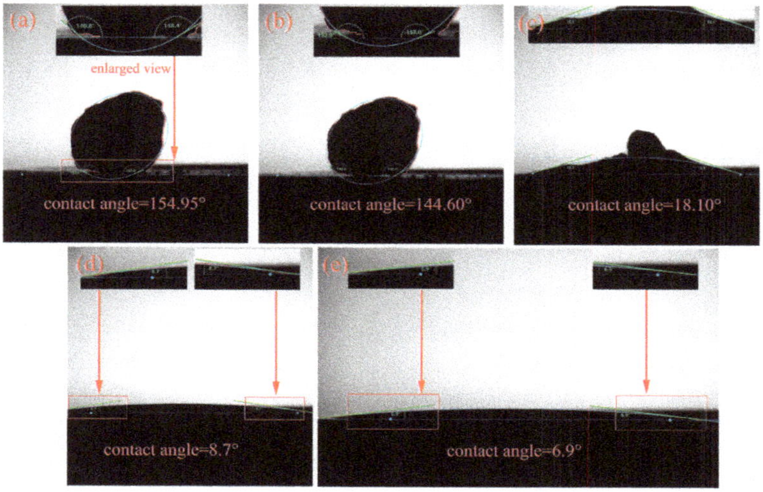

Figure 2. Photographs of the $MgCl_2$-Ti contact angle test: (**a**) 25 °C, (**b**) 714 °C, (**c**) 900 °C, (**d**) holding the temperature at 900 °C for 1 min, and (**e**) holding the temperature at 900 °C for 5 min.

Figure 3 displays the XPS spectra of O1s and Ti2p on the surface of titanium sponge sintered in different mediums. The O1s spectra exhibit two characteristic peaks, with binding energies of 531.0 eV and 531.3 eV, corresponding to the O element in TiO_2 and TiO_2, respectively. Additionally, the peak at 532.8 eV corresponds to the O element in bound water [25]. Thus, when titanium particles are sintered in a molten Mg medium, TiO_2 is present on the surface of sponge titanium. After sintering the titanium particles in a molten Mg:$MgCl_2$ (1:1) medium and a molten $MgCl_2$ medium, the surface of the titanium particles contains TiO_2 and TiO. The Ti2p spectrum can be fitted into three characteristic peaks, where the peaks at binding energies of 458.5 eV and 464.7 eV correspond to Ti^{4+} [26,27] in TiO_2, and the peak at 453.86 eV corresponds to Ti (0) [28]. Following sintering in the three media, the characteristic peaks of TiO_2 and elemental Ti (0) can be observed on the surface of titanium sponge, indicating that oxygen enters the titanium crystal lattice on the surface of the titanium particles and partially forms Ti-O bonds. In other words, there is an oxide layer on the surface of the titanium particles after liquid-phase sintering. This is because molten $MgCl_2$ does not have a reducing effect, while titanium powder is highly reactive. During sintering, both the water introduced by $MgCl_2$ and the brief contact of titanium powder with air will promote the oxidation of sponge titanium, forming TiO_2 and TiO. Molten Mg can reduce the titanium oxides on the surface of sponge titanium, but due to the short reduction time, the reduction is not fully complete, leaving some residual TiO_2.

Figure 4 illustrates the changes in oxygen content of titanium particles after liquid-phase sintering in different mediums. In molten Mg, the oxygen content of titanium particles increases by only 0.11 wt% (growth rate: 12.1%). In the medium of molten Mg:$MgCl_2$ = 1:1, the oxygen content increases by 2.51 wt% (growth rate: 275.8%). In molten $MgCl_2$, the oxygen content increases significantly by 3.90 wt% (growth rate: 428.6%). The oxygen content in titanium particles increases with the increase in molten $MgCl_2$ content. This is because a small amount of water-absorbing $MgCl_2$, when melted, results in the adsorption of MgO on the surface of titanium particles. Furthermore, molten $MgCl_2$ cannot effectively reduce the oxide layer on the surface of titanium particles. On the other hand, molten Mg can help reduce the increase in oxygen content on the surface of titanium particles during sintering. However, MgO, a by-product of the reduction process, is difficult to eliminate from the surface of titanium particles and remains within the particles after distillation. As a result, compared to other mediums, the titanium particles sintered in molten Mg show the smallest increase in oxygen content.

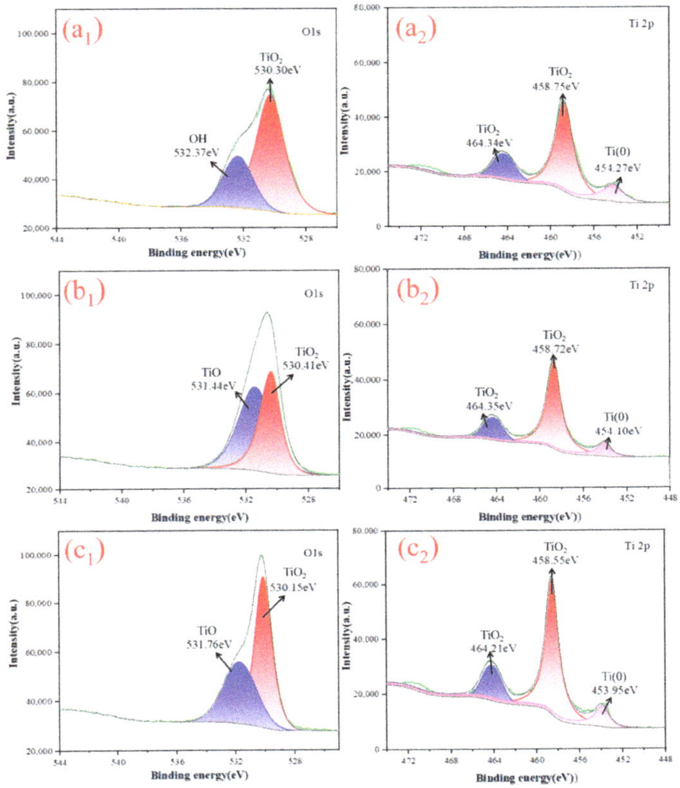

Figure 3. O1s XPS spectra of micron-sized titanium particles sintered at 1100 °C in (**a$_1$**) a molten Mg media, (**b$_1$**) a molten Mg:$MgCl_2$ (1:1) media, and (**c$_1$**) a molten $MgCl_2$ media. Ti2p XPS spectra of micron-sized titanium particles sintered at 1100 °C in (**a$_2$**) a molten Mg media, (**b$_2$**) a molten Mg:$MgCl_2$ (1:1) media, and (**c$_2$**) a molten $MgCl_2$ media.

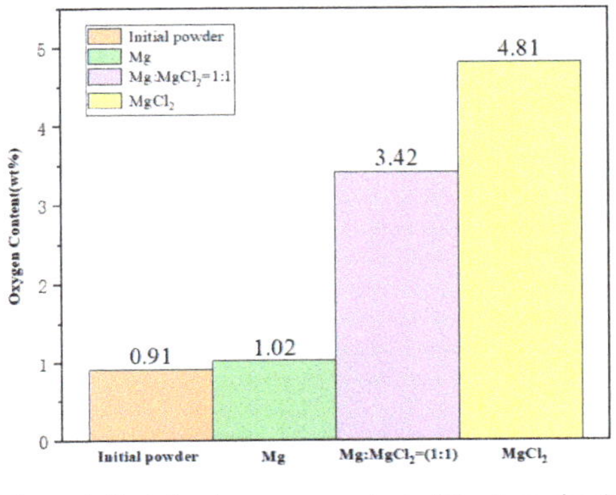

Figure 4. Variation in oxygen content of titanium after 2 h of sintering of titanium particles in different mediums at 1100 °C.

3.2. Aggregation Behavior of Titanium Particles

Figure 5 depicts a metallographic photograph of titanium particles aggregated in a molten $MgCl_2$ medium at different temperatures for 2 h. In the metallographic diagrams, white area represents titanium particles or aggregates, while black area represents pores. As the temperature increases, the wettability between the liquid phase and the solid phase improves [29,30]. The molten medium can spread more thoroughly over the surface of the titanium particles, enhancing the capillary forces between them and accelerating their aggregation. At 800 °C (Figure 5a), the number of titanium particles in the medium per unit area is small and sparsely distributed, accounting for 17.2% of the area. As the temperature increases to 900 °C and 1000 °C (Figure 5b,c), the number of titanium particles in the medium per unit area begins to increase, and the degree of aggregation gradually intensifies. The area proportion occupied by titanium particles reaches 20.0% and 31.4%, respectively. There are still gaps between adjacent titanium particles, indicating that they have not yet fully connected with each other, and the titanium particles remain in the aggregation stage. At 1100 °C (Figure 5d), the diffusion of titanium atoms becomes more pronounced. The area proportion of titanium particles further increases to 35.4%, and adjacent titanium particles become connected, resulting in the aggregation of titanium particles and the formation of a sponge titanium skeleton.

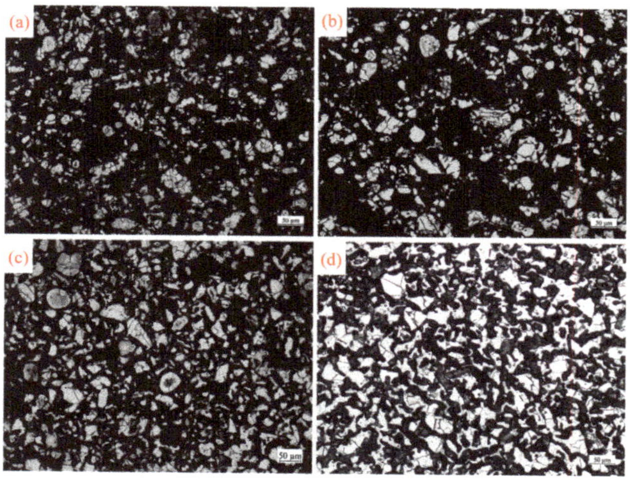

Figure 5. Metallographic photographs of titanium particles after 2 h of aggregating in molten $MgCl_2$ at different temperatures: (**a**) 800 °C, (**b**) 900 °C, (**c**) 1000 °C, and (**d**) 1100 °C.

Figure 6 presents a metallographic photograph of titanium particles aggregated in a molten $Mg{:}MgCl_2$ (1:1) medium at different temperatures for 2 h. The wettability between Mg and Ti is better than that between $MgCl_2$ and Ti. In molten Mg, the capillary force between titanium particles is stronger, accelerating the aggregation speed of titanium particles. Additionally, molten Mg has a reducing effect, which reduces the oxide layer on the surface of titanium particles and weakens the obstacle to the diffusion of titanium atoms. This further accelerates the aggregation of titanium particles, allowing the formation of a sponge titanium skeleton at a lower temperature. Compared to the molten $MgCl_2$ medium, the increased content of molten Mg in the mixed medium accelerates the aggregation rate of titanium particles. At 800 °C (Figure 6a), the proportion of titanium particles is 19.7% (an increase of 2.5% compared to Figure 5a), and their distribution remains relatively dispersed. At 900 °C (Figure 6b), titanium particles continue to gather, and their quantity increases, resulting in an area proportion of titanium particles of 21.4%. Upon reaching 1000 °C (Figure 6c), the area ratio of titanium particles reaches 32.2%. Adjacent titanium particles are connected to each other, leading to the aggregation of titanium particles and the formation of a sponge titanium skeleton.

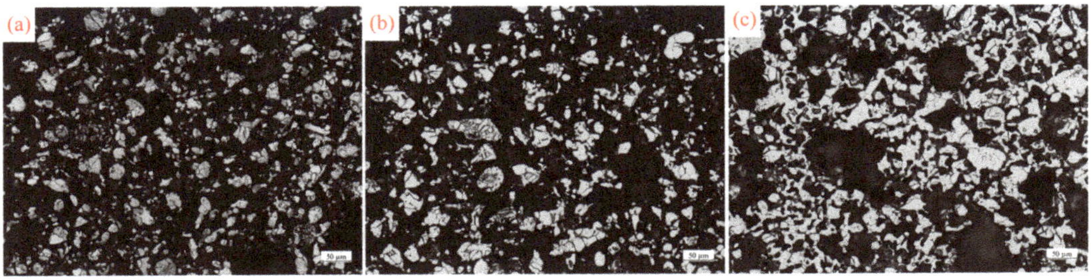

Figure 6. Metallographic photographs of titanium particles after 2 h of aggregating in molten Mg:$MgCl_2$ (1:1) at different temperatures: (**a**) 800 °C, (**b**) 900 °C, and (**c**) 1000 °C.

Figure 7 displays metallographic photographs of titanium particles after liquid-phase aggregating in molten Mg at different temperatures for 2 h. The content of Mg in the molten medium continues to increase, and the wetting and reducing effects of the molten medium are strengthened, which accelerates the aggregation speed of titanium particles. At 800 °C and 900 °C (Figure 7a,b), titanium particles continue to aggregate, accounting for 23.2% and 26.8%, respectively, which is significantly higher compared to the other two media. At 1000 °C (Figure 7c), the aggregation degree of titanium particles increases significantly, resulting in an area proportion of 34.3%. The titanium particles are interconnected, forming an aggregated structure resembling a sponge titanium skeleton. As the content of molten Mg in the medium increases, the aggregation speed of titanium particles also increases, leading to a higher area proportion at the same temperature.

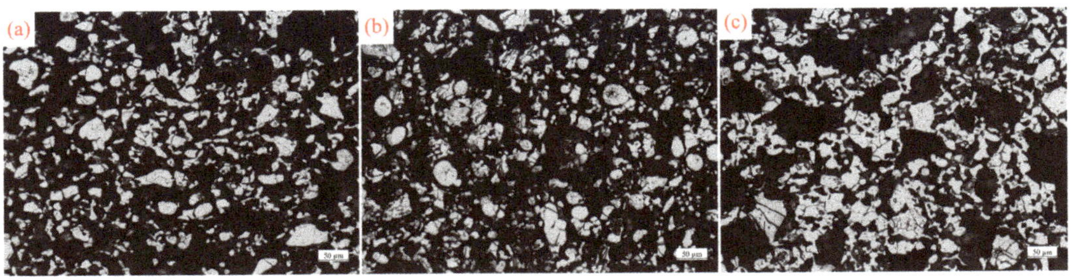

Figure 7. Metallographic photographs of titanium particles after 2 h of aggregating in molten Mg at different temperatures: (**a**) 800 °C, (**b**) 900 °C, and (**c**) 1000 °C.

Figure 8 depicts the micrograph of titanium particles aggregated in different media at 900 °C for 2 h. In the case of aggregating in a molten $MgCl_2$ medium, titanium exhibits no solubility, resulting in the preservation of the original irregular morphology of titanium particles. Fine titanium particles and sharp edges of titanium particles cannot be dissolved, leading to their accumulation on the surface of larger particles (Figure 8a). According to the Ti-Mg binary phase diagram [31,32], titanium possesses certain solubility in molten Mg, which increases gradually with temperature. During aggregating, particles with high surface energy and sharp edges tend to dissolve first [33]. They subsequently nucleate and precipitate on the surface of larger particles or defective areas, causing the disappearance of smaller particles and the growth and spheroidization of larger particles. This process, known as Oswald ripening, can be observed in Figure 8a,b (highlighted by the red boxes). Consequently, the fine particles dissolve and vanish, while the sharp edges of larger particles continuously dissolve and spheroidize. The Oswald ripening phenomenon of titanium particles in molten Mg accelerates the diffusion of titanium atoms through the continuous dissolution and reprecipitation of titanium. This enhances solid-phase mass transfer, accelerating the aggregation of titanium particles. The aggregation of titanium particles leads to the formation of more particle interfaces, as shown in Figure 8c. The microstructure of titanium particles after aggregation is similar to that of mesoporous hematite/alumina nanocomposites [34] and iron oxide nanochains coated with silica [35], as reported by Marin Tadic. The particles contact and adhere to each other to form a 'spongelike' structure, with numerous pores between the particles.

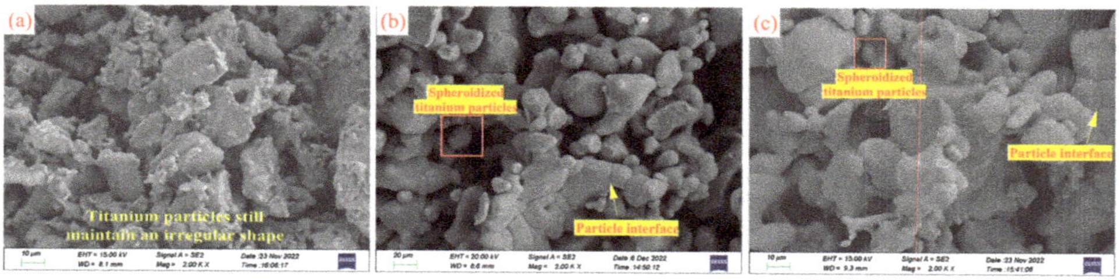

Figure 8. SEM images of titanium particles aggregated in different mediums for 2 h at 900 °C: (**a**) molten $MgCl_2$ medium, (**b**) molten $MgCl_2$:Mg (1:1) medium, and (**c**) molten Mg medium.

Figure 9 illustrates the changes in the average size of titanium particle aggregates with temperature after aggregating in various media for 2 h. The average size of titanium particle aggregates increases with temperature in all three mediums. In the case of aggregating in molten $MgCl_2$, small particles accumulate on the surface of larger particles. As the temperature rises, the number and size of aggregates formed by small particles increase. In the medium containing molten Mg, titanium particles undergo continuous growth through Oswald ripening. Simultaneously, the reduction of the oxide layer on the particle surfaces by molten Mg accelerates mass transfer between titanium particles, facilitating faster particle aggregation and the formation of interfaces. With increasing temperature, the coarsening rate of particles intensifies, resulting in a greater number of particles in contact with each other and larger aggregate sizes.

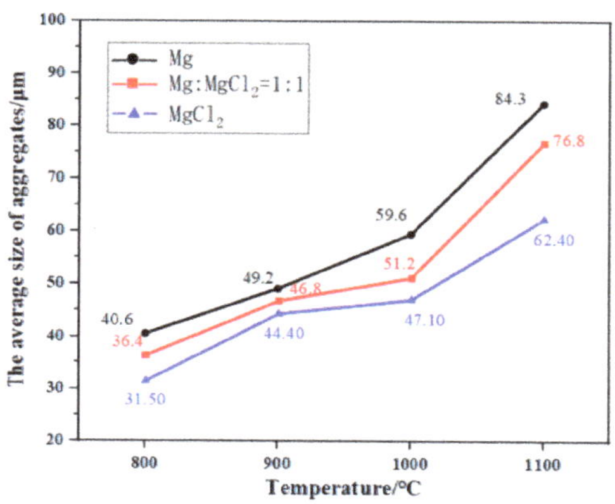

Figure 9. Variation in particle size with temperature after aggregating of titanium particles in different mediums for 2 h.

The average size of aggregates is largest in molten Mg and smallest in molten $MgCl_2$. This is attributed to the formation of a liquid-phase film that covers the surface of titanium particles during aggregating. Capillary forces drive the liquid phase to fill the gaps between particles, promoting particle aggregation, increasing the contact area between particles, enhancing mass transfer, and facilitating easier diffusion of atoms between adjacent titanium particles through the liquid-phase channels. Compared to molten $MgCl_2$, molten Mg exhibits better wetting properties toward titanium and stronger capillary action [36], which promotes particle aggregation and leads to larger aggregate sizes. Additionally, molten Mg reduces the oxide layer on the surface of titanium particles, accelerating mass transfer between particles and promoting particle aggregation. On the other hand, molten $MgCl_2$ cannot reduce the oxide film on the surface of titanium particles, thereby inhibiting titanium atom diffusion and impeding particle aggregation. Therefore, a higher content of molten Mg in the medium facilitates greater particle aggregation and faster growth rates of aggregates.

The process of titanium particle aggregation involves the precipitation of dissolved titanium atoms from molten Mg, the nucleation of titanium atomic clusters to fill the pores, and the nucleation and growth process after particle contact, thereby reducing the porosity of titanium particle aggregates. The nucleation and growth rates of titanium particles during the aggregation process are uniform and occur at a certain rate, and this process, controlled by nucleation and growth dynamics, can be effectively explained by the Johnson–Mehl–Avrami (JMA) model [37–39]. Using the JMA model, the aggregation process of titanium particles at different temperatures and times can be studied to explain the kinetics of the aggregation process. Its expression formula is as follows:

$$\xi = 1 - \exp\left(-kt^{n}\right) \tag{1}$$

where "ξ" presents the percentage of phase proportion, which can be expressed by the relative density of aggregates of titanium particles; "k" presents a kinetics constant dependent on temperature (T); "t" presents time; and "n" presents Avrami index, which reflects kinetics of the aggregation process of titanium particles. The evolution of Equation (1) yields Equation (2), where "$1 - \xi$" can be expressed as the porosity of aggregates of particles.

$$\ln(1/(1 - \xi)) = \ln k + n\ln t \tag{2}$$

Figure 10 illustrates the porosity of titanium particles after aggregation at different temperatures and different times in $MgCl_2$:Mg (1:1). At room temperature, the bulk density of raw titanium powder is 1.26 g/cm^3, and the calculated porosity is 72.0%. After aggregating in $MgCl_2$:Mg (1:1) at 800 °C for 1 h, the porosity of titanium aggregates rapidly decreases to 61.16%. At the same aggregation time, with the increase in temperature, the wettability, reduction, and dissolution–precipitation effect of molten Mg are enhanced, and the diffusion of Ti atoms is also strengthened, which promotes the aggregation process of the titanium particles. After the titanium particles are aggregated at 900 °C for 1 h, the porosity of the particles decreases to 60.61%. The porosity of the aggregates decreases with the increase in temperature and time.

Figure 11 illustrates the fitting curves of the relationship between $\ln(1/(1 - \xi))$ and lnt at different temperatures, which are plotted according to the data in Figure 10. As can be seen from Figure 11, the relationship between $\ln(1/(1 - \xi))$ and lnt basically presents a linear relationship at different temperatures. According to Equation (2), the Avrami index "n" and the kinetics constant "k" can be calculated by the slope and intercept of the fitted lines in the figure, respectively, and their values are shown in Table 1.

According to the kinetics equation, the corresponding kinetics curve is shown in Figure 12. At room temperature, the bulk density of raw titanium powder is 1.26 g/cm^3, and its relative density is 28.0%. In this paper, it is considered that the initial relative density in the kinetics curve is 28.0% (i.e., $\xi = 0.28$). Figure 12a shows that at the same time, the higher the temperature, the higher the density of the titanium particle aggregates and the lower the porosity. This is because as the temperature increases, the diffusion of titanium atoms speeds up. Simultaneously, the wettability, reducibility, and solubility of molten Mg improve, promoting the diffusion of titanium atoms, accelerating particle aggregation, and reducing aggregate porosity. However, when the temperature is low, the activation energy for atomic diffusion is insufficient, and the density of the aggregates remains close to 80% even after 400 h of aggregation. Figure 12b indicates that the porosity of the aggregates increases rapidly from 0 to 3 h and then increases slowly with prolonged time. This is because during the initial 0 to 3 h, molten Mg fully dissolves the titanium in the concave–convex or angular parts of the surface, reaching dynamic equilibrium, while also essentially completing the reduction of the oxide layer on the titanium particle surfaces. Over time, the dissolution–precipitation effect significantly weakens the filling of pores, thereby slowing down the aggregation rate of titanium particles. From the kinetics curves, it can be observed that during the aggregation of titanium particles, increasing the temperature from 800 °C to 900 °C results in only a slight increase in the relative density of the aggregates at the same time. However, with prolonged time, the relative density increases significantly. This indicates that time has a more significant effect on the relative density and porosity of titanium sponge during the aggregation process.

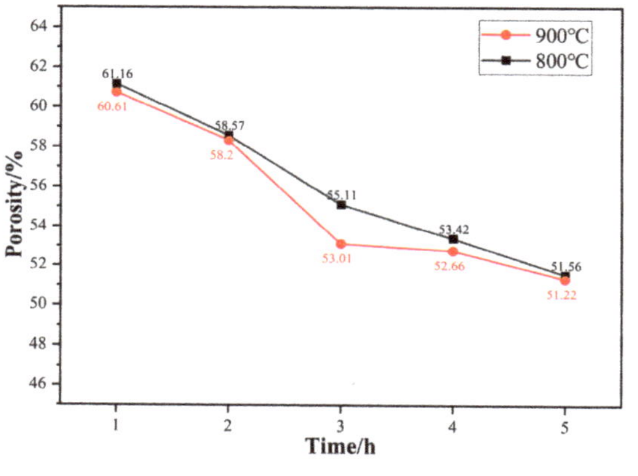

Figure 10. Porosity of titanium particles after aggregation at different temperatures and different times in $MgCl_2$:Mg (1:1).

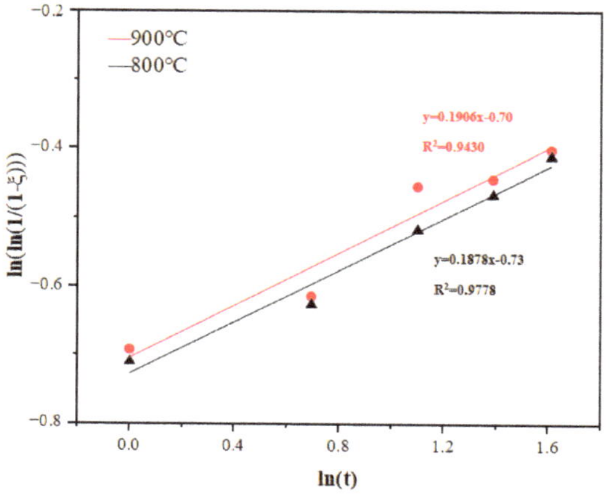

Figure 11. Fitting curves of the relationship between $\ln(1/(1-\xi))$ and lnt at different temperatures.

Table 1. The kinetics equation of the aggregation of titanium particles at different temperatures.

Temperature	Kinetics Formulas	$1-\xi$				
		1 h	2 h	3 h	4 h	5 h
800 °C	$\xi = 1 - \exp(-0.4819t^{0.1878})$	0.6116	0.5857	0.551	0.5342	0.5156
900 °C	$\xi = 1 - \exp(-0.4966t^{0.1906})$	0.6061	0.5820	0.5301	0.5266	0.5122

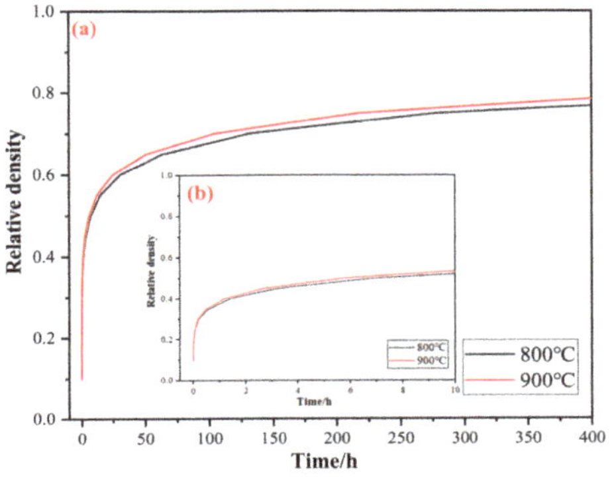

Figure 12. Kinetics curves of titanium particles aggregation at different temperatures in $MgCl_2$:Mg (1:1): (**a**) in the range of 0~400 h and (**b**) partially enlarged detail in the range of 0~10 h.

3.3. Densification Behavior of Sponge Titanium

Figure 13 illustrates the densification behavior of titanium sponge in three different media. In the metallographic diagrams, white area represents the titanium sponge skeleton, while black area represents the pores. At 1100 °C, the wetting, dissolution, and reduction caused by molten Mg accelerate the densification of titanium sponge. As a result, the area ratio of the titanium sponge skeleton is the smallest in molten $MgCl_2$ and the largest in molten Mg, measuring 36.2% (in molten $MgCl_2$, Figure 13(a_1)) and 38.6% (in $MgCl_2$:Mg (1:1), Figure 13(b_1)), respectively. At 1200 °C, compared to the values at 1100 °C, the area proportion of the titanium sponge skeleton increases to 38.9% (in molten $MgCl_2$, Figure 13(a_2)), 47.5% (in $MgCl_2$:Mg (1:1), Figure 13(b_2)), and 52.5% (in molten Mg, Figure 13(c_2)), respectively. At 1300 °C, the area proportion of the titanium sponge skeleton further increases to 52.3% (in molten $MgCl_2$, Figure 13(a_3)), 60.0% (in $MgCl_2$:Mg (1:1), Figure 13(b_3)), and 78.7% (in molten Mg, Figure 13(c_3)).

The wettability and reducibility of molten Mg promote the diffusion of titanium atoms. Simultaneously, titanium dissolves in molten Mg at high temperatures. The precipitation of titanium clusters at the pores reduces the surface energy of the system due to the high surface energy at the pores. This promotes the nucleation and precipitation of titanium atoms at the pores, allowing for local filling of macropores and rapid reduction in the pore area. In molten $MgCl_2$, titanium cannot fill the pores through dissolution–precipitation, hindering pore shrinkage. Consequently, the total pore area of the titanium sponge is larger than that in the molten Mg medium (Figure 13d). At 1100 °C, the dissolution of titanium particles in molten Mg causes their shape to become spheroidized. This change promotes alterations in pore shape and increases the pore size between the skeleton structures, resulting in the largest average pore diameter of the titanium sponge. As the temperature rises, the solubility of titanium in molten Mg increases, leading to stronger precipitation and filling of titanium atoms and a gradual decrease in the average pore size of the titanium sponge. In molten $MgCl_2$, the fine pores in the titanium sponge gradually shrink and disappear, while molten $MgCl_2$ tends to remain in the large pores of the titanium sponge, hindering shrinkage and causing a continuous increase in the average pore size of the titanium sponge (Figure 13e). At 1100 °C, the spheroidization of titanium particles in molten Mg alters the pore shape between the skeleton structures and weakens the mechanical locking between particles, resulting in a high porosity of the titanium sponge. In molten $MgCl_2$, the titanium sponge continuously densifies through atomic diffusion, leading to a continuous reduction in porosity. As the temperature rises, the diffusion of titanium atoms becomes the dominant factor. Compared to molten $MgCl_2$, the titanium sponge exhibits higher density in molten Mg, with the lowest porosity, while the highest porosity is observed in $MgCl_2$.

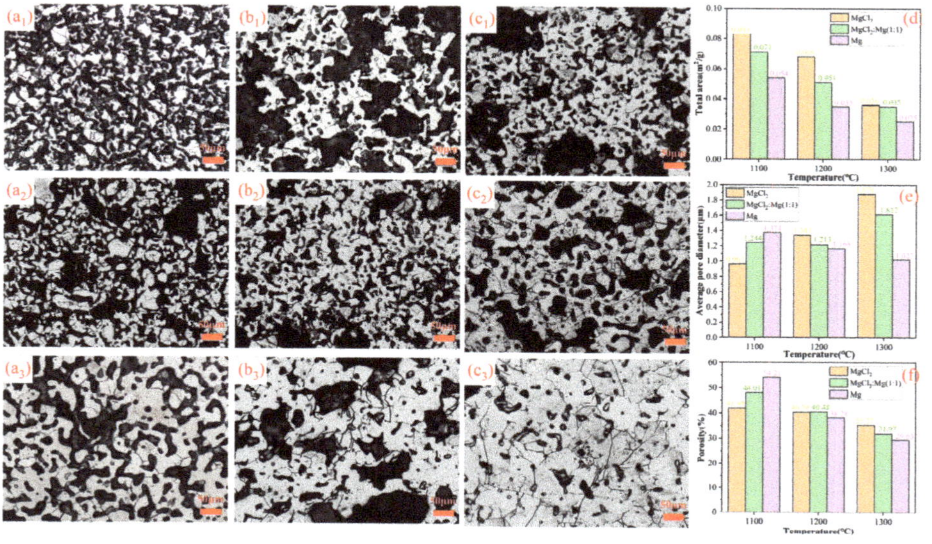

Figure 13. Densification behavior of sponge titanium in three different mediums and temperature: (**a_1–a_3**) represent the metallographic diagrams of sponge titanium densified in molten $MgCl_2$ at 1100 °C, 1200 °C, and 1300 °C, respectively; (**b_1–b_3**) represent the metallographic diagrams of sponge titanium in $MgCl_2$:Mg (1:1) densified at 1100 °C, 1200 °C, and 1300 °C, respectively; (**c_1–c_3**) represent the metallographic diagrams of sponge titanium densified in molten Mg at 1100 °C, 1200 °C, and 1300 °C, respectively; the length of the scale in the photomicrograph represents 50 μm; (**d–f**) represent the variations in total pore area, average pore size, and porosity of the titanium sponge samples.

Figure 14 illustrates the porosity variation of titanium sponge in $MgCl_2$:Mg (1:1) after densification at different temperatures and at different times. With the extension of time, the mass transfer between solid titanium is more sufficient, and the number of titanium clusters precipitated in molten Mg increases, enhancing the filling effect on the pores in the skeleton and continuously reducing the porosity of the sponge titanium. The porosity of titanium sponge decreased from 55.47% to 44.87% at 1000 °C and from 53.91% to 37.31% at 1100 °C. For the same time period, the porosity of titanium sponge at 1100 °C is lower than that at 1000 °C. This is because, at higher temperatures, the diffusion of titanium atoms and grain growth occur more rapidly, promoting the densification of the titanium sponge. Additionally, the strengthened dissolution–precipitation effect of molten Mg on titanium increases the filling effect on the pores, leading to a reduction in the porosity of the titanium sponge skeleton.

The Avrami formula is also used to study the kinetics of sponge titanium skeleton densification. The "ξ" in Equation (1) can be used to represent the relative density of the titanium sponge skeleton. The curves for $\ln(1/(1-\xi))$ and lnt are fitted by using the same method as in Section 3.2 and are shown in Figure 15. As can be seen from Figure 15, the relationship between $\ln(1/(1-\xi))$ and lnt basically presents a linear relationship at different temperatures. According to Equation (2), the Avrami index "n" and the kinetics constant "k" are calculated and listed in Table 2.

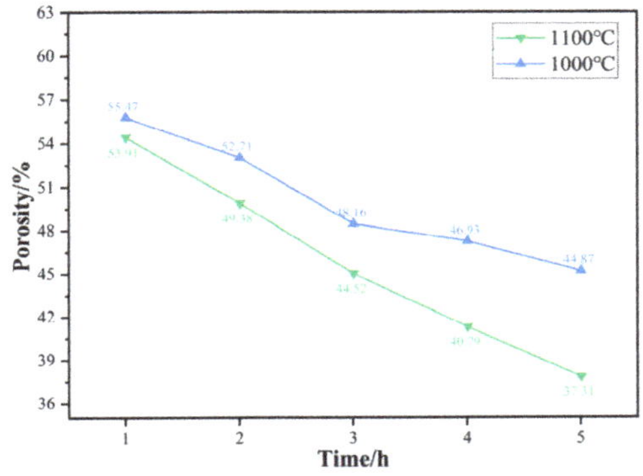

Figure 14. Porosity of titanium sponge after densification at different temperatures and different times in $MgCl_2$:Mg (1:1).

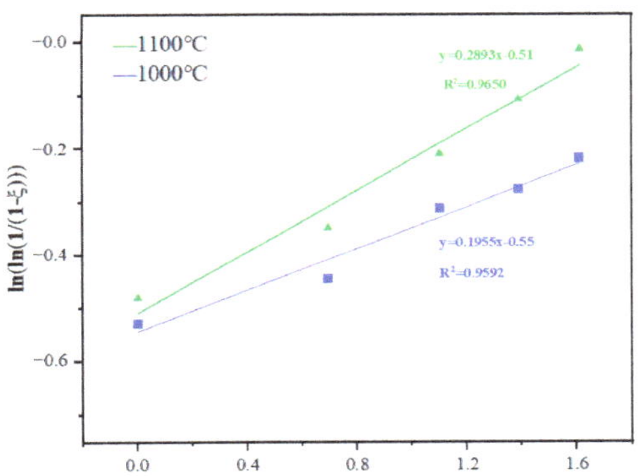

Figure 15. Fitting curves of the relationship between $\ln(1/(1-\xi))$ and lnt at different temperatures.

Table 2. Kinetics formulas at different temperatures.

Temperature	Kinetics Formulas	$1-\xi$				
		1 h	2 h	3 h	4 h	5 h
1000 °C	$\xi = 1 - \exp(-0.5769t^{0.1955})$	0.5547	0.5271	0.4816	0.4693	0.4487
1100 °C	$\xi = 1 - \exp(-0.6005t^{0.2893})$	0.5391	0.4938	0.4452	0.4079	0.3731

Figure 16 shows the kinetics curve of the titanium sponge skeleton at different temperatures in $MgCl_2$:Mg (1:1). The kinetics curve exhibits a rapid rise from 0 h to 3 h, and then shows a slow increase from 3 h to 80 h and continues to extend the time. The kinetics curve essentially remains stable, indicating that the densification process has reached its limit. In the early stage of densification, there are a lot of pores in the titanium sponge. Through the diffusion of titanium atoms and the filling of the precipitated titanium atoms, the pores in the titanium sponge shrink and close rapidly, leading to a rapid decrease in porosity. In the latter stage of densification, as the number of pores decreases, it becomes difficult for the remaining pores to continue shrinking without external force, resulting in a slower decrease in porosity. With the increase in temperature, the diffusion of titanium atoms is accelerated, and the solid-phase mass transfer rate of Ti is increased, resulting in the increase in the limit value of the kinetics curve, indicating that the maximum degree of densification is increased.

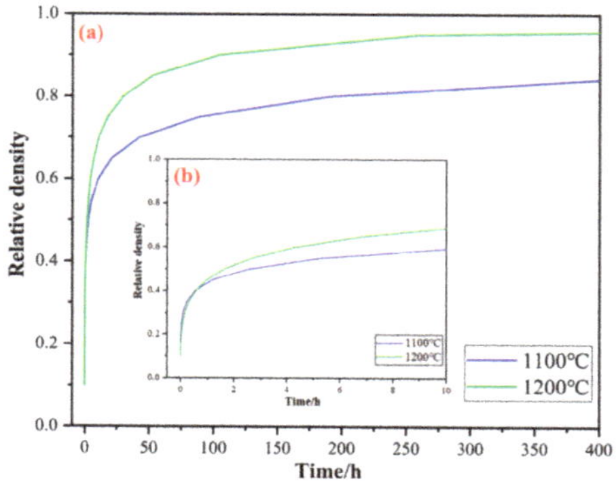

Figure 16. Kinetics curve of the titanium sponge skeleton densification at different temperatures in $MgCl_2$:Mg (1:1): (**a**) in the range of 0~400 h and (**b**) partially enlarged detail in the range of 0~10 h.

From the kinetics curves of the densification process, it is evident that the trend of relative density increase in the sponge titanium is similar over time. But at 1200 °C, the sponge titanium exhibits a significantly higher relative density. This shows that during the densification stage of the sponge titanium, temperature is the main factor affecting its relative density and porosity.

4. Conclusions

This paper investigates the effects of the composition of a high-temperature liquid medium (the molten Mg-$MgCl_2$ system) on the aggregation–growth and densification behavior of titanium particles, coming to the following conclusions:

1. Compared to molten $MgCl_2$, molten Mg demonstrates superior reduction, dissolution, and wetting effects, which promote the diffusion of titanium atoms, facilitate the aggregation of titanium particles, and accelerate the formation of the titanium sponge skeleton. In the molten $MgCl_2$ medium, titanium particles aggregate and grow at 1100 °C, resulting in the formation of the titanium sponge skeleton. In the molten Mg medium and $MgCl_2$:Mg (1:1) medium, titanium particles aggregate and grow at 1000 °C, leading to the formation of the titanium sponge skeleton. With the increase in temperature, the wettability, reducibility, and solubility of the molten medium are improved, promoting the diffusion of titanium atoms, strengthening the capillary forces between titanium particles, and accelerating the aggregation of titanium particles.
2. The Oswald ripening mechanism promotes the coarsening of titanium particles, and the reduction in molten Mg accelerates the aggregation of particles, causing the aggregates to grow rapidly in molten Mg. When aggregated in molten $MgCl_2$, the aggregate size increases through the gathering of small particles on the surface of large particles, resulting in a slower growth rate.
3. With an increase in the relative content of molten Mg in the medium, the diffusion of titanium atoms is promoted, and the solid titanium's mass transfer is accelerated, which benefits the densification of the titanium sponge. At different temperatures, the area ratio of the titanium sponge skeleton is highest in molten Mg, while the total pore area, average pore size, and porosity are lower compared to the other media.
4. Kinetics studies show that the porosity of titanium particle aggregates and the porosity of the titanium sponge skeleton decrease with increasing temperature at the same time, and the limiting value of the relative density increases. The aggregation rate of titanium particles and the densification rate of titanium sponge were higher at 0~3 h. From 3 h to 80 h, the aggregation rate of titanium particles and the densification rate of titanium sponge gradually decreased and continued to extend the time. The

relative density basically remained unchanged, and the density of aggregates and titanium sponge reached the limit.

In summary, increasing the Mg content in the melting medium and raising the temperature will accelerate the aggregation–growth and densification of titanium particles, resulting in low-quality sponge titanium with low porosity and high compactness. Therefore, in industrial production, the aggregation–growth rate and densification degree of titanium particles can be delayed by controlling the temperature (enhancing heat dissipation, etc.) and adjusting the composition of the melting medium (reducing the emission of $MgCl_2$, etc.) to obtain high-quality titanium sponge with a loose structure.

Author Contributions: Conceptualization, Y.L. and K.L.; investigation, K.L. and Z.S.; data curation, X.Y.; writing—original draft, X.Y.; writing—review and editing, J.L. and K.L.; funding acquisition, K.L. and J.L. All authors have read and agreed to the published version of the manuscript.

Funding: This work is supported by the National Nature Science Foundation of China (Fund number: 52341402) and the Science and Technology Department of Sichuan Province (Fund number: 2023ZDZX0027).

Institutional Review Board Statement: Not applicable.

Informed Consent Statement: Not applicable.

Data Availability Statement: The original contributions presented in the study are included in the article, further inquiries can be directed to the corresponding author.

Conflicts of Interest: Authors Kaihua Li and Zhuo Sheng were employed by the company Pangang Group Research Institute Co., Ltd. The remaining authors declare that the research was conducted in the absence of any commercial or financial relationships that could be construed as a potential conflict of interest.

References

1. Lutjering, G.; Williams, J.C. *Titanium*; John Wiley & Sons, Ltd.: Hoboken, NJ, USA, 2007.
2. Kishimoto, A.; Uda, T. In-situ observation on the magnesiothermic reduction of $TiCl_4$ around 800 °C by microfocus X-ray fluoroscopy. *J. Alloys Compd.* **2021**, *874*, 159855. [CrossRef]
3. Nagesh, C.R.; Rao, C.S.; Ballal, N.B.; Rao, P.K. Mechanism of titanium sponge formation in the Kroll reduction reactor. *Metall. Mater. Trans. B* **2004**, *35*, 65–74. [CrossRef]
4. Wartman, F.S.; Baker, D.H.; Nettle, J.R.; Homme, V.E. Some observations on the Kroll process for titanium. *J. Electrochem. Soc.* **1954**, *101*, 507. [CrossRef]
5. Wang, W.; Wu, F.; Jin, H. Enhancement and performance evaluation for heat transfer of air cooling zone for reduction system of sponge titanium. *Heat Mass Transf.* **2017**, *53*, 465–473. [CrossRef]
6. Evdokimov, V.I.; Krenev, V.A. Magnesium reduction of titanium tetrachloride. *Inorg. Mater.* **2002**, *38*, 490–493. [CrossRef]
7. Noda, T. Outline of Presentation on Development in Ti Sponge Production in Halda Century. *Titan. Jpn.* **1994**, *42*, 85–89.
8. Okabe, T.H.; Sadoway, D.R. Metallothermic reduction as an electronically mediated reaction. *J. Mater. Res.* **1998**, *13*, 3372–3377. [CrossRef]
9. Uda, T.; Okabe, T.; Kasai, E.; Waseda, Y. Direct evidence of electronically mediated reaction during $TiCl_4$ reduction by magnesium; $TiCl_4$ no magnesium kangen ni okeru dendotai wo kaishita hanno (EMR) no jissho. *J. Jpn. Inst. Met. Mater.* **1997**, *61*, 602–609. [CrossRef]
10. Shimosaki, S.; Kuramoto, M. Production rate of titanium by reduction of titanium tetrachloride with magnesium; $TiCl_4$ no Mg kangen ni yoru kinzoku Ti no seisei sokudo. *Kagaku Kogaku Ronbunshu* **1995**, *21*, 740–745. [CrossRef]
11. Uda, T.; Okabe, T.H.; Waseda, Y. Location control of titanium deposition during magnesiothermic reduction of $TiCl_4$ by long range electronically mediated reaction. *J. Jpn. Inst. Met. Mater.* **1998**, *62*, 76–84. [CrossRef]
12. Nagesh, C.R.V.S.; Ramachandran, C.S.; Subramanyam, R.B. Methods of titanium sponge production. *Trans. Indian. Inst. Met.* **2008**, *61*, 341–348. [CrossRef]
13. Zhou, L.G. *Research on Key Technologies of Magnesium Reduction in Titanium Sponge Production*; Yunnan University: Kunming, China, 2016.
14. Wang, C. *Cause Analysis and Control Measures of Titanium Sponge Densification*; Kunming University of Science and Technology: Kunming, China, 2015.
15. Lee, J.C.; Jung, J.Y.; Sohn, H. Effect of $TiCl_4$ Feeding Rate on the Formation of Titanium Sponge in the Kroll Process. *J. Korean Inst. Met. Mater.* **2012**, *50*, 745–751.

16. Hockaday, S.A.C.; Bisaka, K. Experience and results from running of a 1 kg Ti scale Kroll reactor. In *Proceedings of the Advanced Metals Initiative: Light Metals Conference 2010*; Southern African Institute of Mining and Metallurgy: Johannesburg, South Africa, 2010; pp. 265–280.
17. Withers, J.; Laughlin, J.; Elkadi, Y.; DeSilva, J.; Loutfy, R.O. A Continuous Process to Produce Titanium Utilizing Metallothermic Chemistry. *Key Eng. Mater.* **2010**, *436*, 55–60. [CrossRef]
18. Bae, H.N.; Kim, S.H.; Lee, G.G.; Jo, S.K.; Jung, J.Y. Study of Thermal Behavior in a Kroll Reactor for the Optimization of Ti Sponge Production. *Mater. Sci. Forum* **2010**, *654–656*, 839–842. [CrossRef]
19. Wang, W.; Wu, F.; Jin, H.; Gao, C. Heat Transfer Model of Molten Bath in Sponge Titanium Production with Magnesiothermic Reduction. *Nonferrous Met. (Extr. Metall.)* **2013**, *11*, 19–21.
20. Dou, S.; Wu, F.; Gao, C. Study on Forced Heat Dissipation for Titanium Sponge Production through Magnesium Thermal Reduction Process. *Nonferrous Met. (Extr. Metall.)* **2013**, *6*, 22–25.
21. Chen, Y.; Li, Z.; Zhang, J.; Xie, L. Effect of Reduction Temperature on Structure of Titanium Sponge Produced by Kroll Process. *Nonferrous Met. (Extr. Metall.)* **2014**, *4*, 29–32.
22. He, Y.; Chen, D.; Sun, X.; Liu, H.; Chen, Y.; Zhou, Y.; Zhao, Y. Simulation of Temperature Field in a 13 t Large-scale Titanium Sponge Reduction Reactor. *Spec. Cast. Nonferrous Alloys* **2021**, *41*, 956–960.
23. Feng, T.; Zheng, W.; Chen, W.; Shi, Y.; Fu, Y.Q. Enhanced interfacial wettability and mechanical properties of Ni@Al_2O_3/Cu ceramic matrix composites using spark plasma sintering of Ni coated Al_2O_3 powders. *Vac. Technol. Appl. Ion Phys. Int. J. Abstr. Serv. Vac. Sci. Technol.* **2021**, *184*, 109938. [CrossRef]
24. Ebrahimi, S.; Heydari, M.S.; Baharvandi, H.R.; Ehsani, N. Effect of iron on the wetting, sintering ability, and the physical and mechanical properties of boron carbide composites: A review. *Int. J. Refract. Met. Hard Mater.* **2016**, *57*, 78–92. [CrossRef]
25. Zhou, R.; Wei, D.; Cheng, S.; Li, B.; Wang, Y.; Jia, D.; Zhou, Y.; Guo, H. The structure and in vitro apatite formation ability of porous titanium covered bioactive microarc oxidized TiO_2-based coatings containing Si, Na and Ca. *Ceram. Int.* **2014**, *40*, 501–509. [CrossRef]
26. González-Elipe, A.R.; Munuera, G.; Espinos, J.P.; Sanz, J.M. Compositional changes induced by 3.5 keV Ar+ ion bombardment in Ni-Ti oxide systems: A comparative study. *Surf. Sci.* **1989**, *220*, 368–380. [CrossRef]
27. Sodhi, R.N.S.; Weninger, A.; Davies, J.E.; Sreenivas, K. X-ray photoelectron spectroscopic comparison of sputtered Ti, Ti6Al4V, and passivated bulk metals for use in cell culture techniques. *J. Vac. Sci. Technol. A Vac. Surf. Film.* **1991**, *9*, 1329–1333. [CrossRef]
28. Biesinger, M.C.; Lau, L.W.M.; Gerson, A.R.; Smart, R.S.C. Resolving surface chemical states in XPS analysis of first row transition metals, oxides and hydroxides: Sc, Ti, V, Cu and Zn. *Appl. Surf. Sci.* **2010**, *257*, 887–898. [CrossRef]
29. Zhang, M.; Chen, C.; Li, C. Wetting and sealing of the interface between silicate glass and copper. *Int. J. Mater. Res.* **2019**, *110*, 163–173. [CrossRef]
30. Fima, P.; Gancarz, T.; Pstruś, J.; Sypień, A. Wetting of Sn-Zn-x In (x = 0.5, 1.0, 1.5 wt%) Alloys on Cu and Ni Substrates. *J. Mater. Eng. Perform.* **2012**, *21*, 595–598. [CrossRef]
31. Murray, J.L. The Mg−Ti (magnesium-titanium) system. *Bull. Alloy Phase Diagr.* **1986**, *7*, 245–248. [CrossRef]
32. Aust, K.T.; Pidgeon, L.M. Solubility of titanium in liquid magnesium. *JOM* **1949**, *1*, 585–587. [CrossRef]
33. Guo, S. *Powder Sintering Theory*; Metallurgical Industry Press: Beijing, China, 1998.
34. Tadic, M.; Panjan, M.; Tadic, B.V.; Kralj, S.; Lazovic, J. Magnetic properties of mesoporous hematite/alumina nanocomposite and evaluation for biomedical applications. *Ceram. Int.* **2022**, *48*, 10004–10014. [CrossRef]
35. Tadic, M.; Kralj, S.; Lalatonne, Y.; Motte, L. Iron oxide nanochains coated with silica: Synthesis, surface effects and magnetic properties. *Appl. Surf. Sci.* **2019**, *476*, 641–646. [CrossRef]
36. Yan, X.; Zhang, Y.; Wang, Q.; Zhang, X.; He, S. Wettability between Titanium-Containing Steels and TiC Ceramic Substrate. *Steel Res. Int.* **2023**, *94*, 2300190. [CrossRef]
37. Liu, X.; Hu, L. Desorption–Recombination kinetics and Mechanism of a Mechanically Milled Nano-Structured $Nd_{16}Fe_{76}B_8$ Alloy During Vacuum Annealing. *J. Alloys Compd.* **2013**, *558*, 142–149. [CrossRef]
38. Jeangros, Q.; Hansen, T.W.; Wagner, J.B.; Damsgaard, C.D.; Dunin-Borkowski, R.E.; Hébert, C.; Van Herle, J.; Hessler-Wyser, A. Reduction of nickel oxide particles by hydrogen studied in an environmental TEM. *J. Mater. Sci.* **2013**, *48*, 2893–2907. [CrossRef]
39. Duan, C.W.; Hu, L.X.; Sun, Y.; Zhou, H.P.; Yu, H. Reaction kinetics for the solid state synthesis of the $AlH_3/MgCl_2$ nano-composite by mechanical milling. *Phys. Chem. Chem. Phys.* **2015**, *17*, 22152–22159. [CrossRef] [PubMed]

MDPI AG
Grosspeteranlage 5
4052 Basel
Switzerland
Tel.: +41 61 683 77 34

Materials Editorial Office
E-mail: materials@mdpi.com
www.mdpi.com/journal/materials

www.ingramcontent.com/pod-product-compliance
Lightning Source LLC
Chambersburg PA
CBHW062036220926
51850CB00010B/2571
* 9 7 8 3 7 2 5 8 4 0 7 0 0 *